Digging for Truth
Between Genesis and the Jurassic

HIS STORY for US
PUBLISHING

Did God Create T-Rex

ARTHUR A.TIGER

Did God Create T-Rex? Between Genesis and the Jurassic

By Arthur A.Tiger.

This balanced exploration of how Christians understand dinosaurs presents young-earth, old-earth, intelligent design, and theistic evolutionary perspectives with fairness and respect. Written for adult Christians interested in science-faith dialogue, parents facing children's questions, educators, seminary students, and pastors, the book provides thoughtful resources for navigating one of the most challenging intersections of biblical faith and scientific discovery.

Published by His Story for Us

www.arthurtigerbooks.com

First Edition: 2025

ISBN 978-2-7332-0292-0

Preface

I must confess that the subject of dinosaurs has captivated me since childhood. I'm certain that virtually every child goes through a dinosaur phase—and for good reason. Here was a fantasy world that actually existed on our Earth. Dragons were real. Monsters walked the land. And unlike fairy tales, we have their bones to prove it.

This is perhaps the only book I've wanted to write since I was a small boy, from the very moment I first saw a real dinosaur. Yes—it was real! I was five years old when my father took me to see "The Legend of the Dinosaur" at our local cinema. I can still remember the terror and awe I felt before those giants on the screen. My hands gripped the armrests, my eyes widened in wonder, and something deep inside me awakened—a lifelong fascination with these "terrible lizards" and what they mean for our understanding of God's creation.

I owe an immeasurable debt to my father for that day. Not just for taking a five-year-old to what was probably an age-inappropriate film, but for nurturing in me the same inquisitive mind that drove him. He was a man who looked at the universe and asked "How?" and "Why?" He saw no conflict between his faith and his curiosity. Questions were not threats to belief but pathways to deeper understanding.

Writing this book has been a journey of discovery—and considerable struggle. I've attempted to be scrupulously objective, to present different Christian perspectives on dinosaurs fairly and thoroughly. Yet I know that complete objectivity is impossible when dealing with questions that touch the very foundations of our faith and our understanding of the world.

In the process of writing and endless battles with editors (who pushed me to be clearer, fairer, and more precise), I've come to realize that this book will satisfy no one completely. Some religious readers will find it too accommodating to science, too willing to entertain old-earth scenarios or evolutionary processes. Some scientifically-minded readers will find it too credulous, too willing to consider young-earth arguments or supernatural interventions.

But perhaps that's as it should be. The intersection of dinosaurs and faith is not a place for easy answers or comfortable certainties. It's a place where our finite minds grapple with infinite mysteries, where bones in the ground challenge us to think deeply about the God who grounds all existence.

This book is ultimately a tribute to my father and his questing spirit. He taught me that faith and curiosity could walk hand in hand, that wondering about dinosaurs didn't diminish worship but could enhance it. He showed me that the same God who inspired Scripture also inscribed His signature in stone through fossils. Most importantly, he demonstrated that honest questions deserve honest exploration, even when—especially when—the answers prove elusive.

I offer this book not as the final word on dinosaurs and faith, but as an invitation to join the conversation. Whether you're a parent facing your child's dinosaur questions, a student wrestling with competing worldviews, or simply someone who's wondered how Tyrannosaurus rex fits into God's plan, I hope you'll find here both information and inspiration for your own journey.

May you approach these ancient creatures with the wonder of a five-year-old and the wisdom of a seeker. May you find in their fossilized remains not a challenge to faith but a call to deeper trust in the God who creates wonders beyond our imagination. And may you, like my father, never lose that precious gift of curiosity that drives us to explore the marvelous works of our Creator.

To Dad, who taught me to look up at the stars and down at the stones, and to see God's glory in both.

A note on approach: Throughout this book, I present multiple Christian perspectives on how to understand dinosaurs in light of biblical faith. While I have my own convictions, I've tried to represent each viewpoint fairly and respectfully. Where I've failed in that attempt, the fault is mine alone. My goal is not to convince you of a particular position but to equip you to think carefully and faithfully about these questions for yourself.

Disclaimer

This book presents multiple Christian perspectives on how dinosaurs relate to biblical faith. While I have striven for accuracy and fairness in representing different viewpoints, readers should understand several important points:

Regarding Scientific Information: All scientific data presented reflects current mainstream paleontological understanding as of the time of writing. Scientific knowledge continually advances, and new discoveries may modify or overturn current theories. Readers are encouraged to consult recent peer-reviewed sources for the latest findings.

Regarding Theological Positions: The young-earth creationist, old-earth creationist, intelligent design, and theistic evolutionary positions presented here are simplified summaries of complex theological and philosophical systems. Each position has extensive literature, internal debates, and nuanced variations that could not be fully explored in a single volume. I have attempted to present the strongest arguments for each position rather than constructing straw men.

Regarding Historical Narratives: Some scenes in this book are narrative reconstructions based on historical facts. For example, while we know William Buckland discovered Megalosaurus and held certain theological views, specific dialogues and personal moments are imagined to bring historical figures to life. Where I have taken such literary license, the core facts remain accurate to the best of my historical knowledge.

Regarding Personal Conviction: While I have my own convictions about these matters, I have attempted to present each position in a way that its thoughtful adherents would recognize as fair. My goal is not to convince readers of a particular position but to provide resources for thinking through these issues carefully and faithfully.

Regarding Salvation: Nothing in this book should be construed to suggest that one's view on the age of the earth or the means of creation affects one's salvation. Historic Christian faith across all major traditions has always affirmed salvation by grace through faith in Jesus Christ, not by holding particular positions on secondary issues.

Regarding Ongoing Debates: The intersection of paleontology and Christian faith remains an area of active discussion and disagreement among sincere believers. This book represents a snapshot of current debates, not a final resolution. Readers are encouraged to continue studying, praying, and discussing these matters with humility and charity.

Regarding Use in Education: Teachers and group leaders using this book should note that it presents multiple viewpoints without endorsing one as definitively correct. It is designed to foster thoughtful discussion rather than provide simple answers.

A Final Note: If this book helps you think more clearly about dinosaurs and faith, I rejoice. If it raises more questions than it answers, that too fulfills its purpose. The goal is not certainty about every detail but confidence in the God who created both ancient dragons and modern disciples, who reveals Himself in Scripture and nature, and who invites us into the adventure of discovery.

May you read with wisdom, disagree with charity, and wonder with worship.

Chapter 1:

When Christians Discovered Dragons

"How many are your works, Lord! In wisdom you made them all; the earth is full of your creatures." — Psalm 104:24

The Girl Who Found Sea Dragons

Lyme Regis, England, 1811. The winter storm had been fierce, one of those tempests that locals said could "wake the dead." Twelve-year-old Mary Anning knew better—storms didn't wake the dead, but they certainly revealed them.

She scrambled across the rocky beach with her brother Joseph, their eyes scanning the freshly exposed cliff face. Mary's fingers were already numb from the cold, but she didn't care. Father had taught them well before his death: after a storm comes opportunity. The cliffs gave up their secrets reluctantly, and only to those who searched with patience and sharp eyes.

"Mary! Over here!" Joseph's voice carried over the crash of waves.

What Mary saw next would change not just her life, but the entire world's understanding of God's creation. Embedded in the cliff face was a skull unlike anything in their collection—four feet long, with eye sockets the size of saucers and rows of sharp, conical teeth. This was no ordinary fossil fish or shell. This was a monster.

Or rather, as the world would soon learn, this was an ichthyosaur—a "fish lizard," one of the first complete specimens ever discovered.

That evening, Mary sat by candlelight reading from the family

Bible, as was her custom. The words of Job seemed to leap from the page: "But ask the beasts, and they will teach you; the birds of the heavens, and they will tell you; or the bushes of the earth, and they will teach you; and the fish of the sea will declare to you" (Job 12:7-8).

What declaration was this ancient sea dragon making? Mary couldn't know that her discovery would ignite debates that rage to this day. She was simply a devoted Christian girl who had found something wonderful and terrible in the rocks—a messenger from a world that existed long before humans walked the earth.

The First Christian Paleontologists

The story of how Christians discovered and named dinosaurs is far more nuanced than many realize. Rather than opposing these discoveries, many early paleontologists were devout believers who saw their work as revealing God's handiwork. Their faith didn't hinder their science—it inspired it.

William Buckland was an unlikely revolutionary. As Dean of Westminster and Oxford's first professor of geology, he balanced his time between Sunday sermons and studying stones. Known for his eccentricities—he claimed to have eaten his way through the animal kingdom, from mice to crocodile—Buckland approached both faith and fossils with infectious enthusiasm.

In 1824, Buckland made history. In a presentation to the Geological Society of London, he described a collection of fossilized bones from a Stonesfield slate quarry. The creature they belonged to was clearly reptilian but of enormous size. Buckland named it Megalosaurus, meaning "great lizard."

"The word of God and the works of God cannot contradict each other," Buckland often said. For him, studying Megalosaurus was an act of worship. Each bone revealed something about the Creator's power and creativity. Where others might see conflict between geological time and Genesis, Buckland saw an opportunity to understand God's methods more deeply.

His approach was typical of the era's "natural theology"—the idea that nature was God's "second book," complementing Scripture. Megalosaurus wasn't a threat to faith; it was another chapter in understanding divine wisdom.

While Buckland worked from his Oxford position, Gideon Mantell pursued paleontology as a country doctor in Sussex. The story of his greatest discovery has become legend, though like many legends, the details vary with each telling.

In 1822, Mantell (or possibly his wife, Mary Ann) discovered unusual teeth in a pile of road gravel near Cuckfield. They resembled the teeth of an iguana but were far too large. Mantell spent years trying to convince the scientific establishment that these teeth belonged to an ancient, gigantic herbivorous reptile.

Mantell's faith sustained him through professional rejection and personal tragedy. He saw geology as revealing "God's grandest works" and wrote extensively about harmonizing Scripture with geological findings. His Iguanodon ("iguana tooth") became the second dinosaur scientifically described.

"There is nothing in geological discovery contrary to revealed religion," Mantell insisted. "The Earth's crust is a great stone book, written by the finger of God."

If Buckland discovered Megalosaurus and Mantell described Iguanodon, it was Richard Owen who realized these creatures represented something entirely new. In 1842, presenting to the British Association for the Advancement of Science, Owen coined a word that would capture imaginations forever: Dinosauria—"terrible lizards."

Owen identified three genera as founding members of this new group: Megalosaurus, Iguanodon, and Hylaeosaurus (discovered by Mantell in 1833). He noted their unique anatomical features:

upright limbs, complex vertebrae, and enormous size. These weren't just big lizards—they were a distinct group of creatures unlike anything alive today.

A devout Anglican, Owen saw dinosaurs as evidence of divine design. He became one of Darwin's fiercest critics, not because he opposed science, but because he believed evolution couldn't explain the complexity he observed. For Owen, each dinosaur bone testified to intentional creation.

Owen's vision extended beyond research. He spearheaded the creation of London's Natural History Museum, designing it as a "cathedral to nature" where visitors could worship God through His works. The museum's architecture, with its soaring arches and intricate decorations depicting creatures both living and extinct, embodied Owen's theology: nature was sacred, worthy of both study and reverence.

Early Church Reactions: From Wonder to Worry

The early 19th century was a unique moment in Christian thought. The Bridgewater Treatises, commissioned to demonstrate "the Power, Wisdom, and Goodness of God, as manifested in the Creation," exemplified the era's confidence. Science and faith weren't competitors—they were partners in revealing truth.

When dinosaurs entered the scene, many Christians received them enthusiastically. Here was proof of God's creative power beyond imagination! The intricate design of a Megalosaurus hip joint or an Iguanodon tooth showed divine craftsmanship at scales previously unimagined.

Sermons of the period often referenced these discoveries. One minister proclaimed, "If God lavished such care on creatures destined for extinction, how much more does He care for humanity, made in His image?" Dinosaurs became arguments for faith, not against it.

However, challenges emerged. Traditional Biblical chronology, calculated from genealogies, suggested Earth was roughly 6,000 years old. But geologists spoke of millions of years. How could these timescales reconcile?

Additionally, extinction itself posed theological problems. If God's creation was "very good," why did He create species destined to disappear? The notion of a perfect creation seemed at odds with a fossil record full of dead ends.

Some attempted harmonization through various theories. The Gap Theory proposed that eons passed between Genesis 1:1 and 1:2. The Day-Age Theory suggested that creation "days" represented geological epochs. Others wondered if Noah's flood might have been regional rather than global. These interpretations allowed Christians to accept geological findings while maintaining Biblical authority. But not everyone was comfortable with such accommodations.

Charles Darwin's On the Origin of Species (1859) transformed the conversation entirely. Suddenly, dinosaurs weren't just extinct creatures—they were evidence for evolution. The same fossils that once proclaimed God's creativity now seemed to support naturalistic explanations.

The great debate between Owen and Thomas Henry Huxley (Darwin's "bulldog") symbolized the growing divide. Both examined the same bones but reached opposite conclusions. Owen saw divine design; Huxley saw natural selection.

Churches began to split along similar lines. Some took a conservative approach, defending traditional interpretations against scientific challenges. Others adopted a more liberal stance, reinterpreting Scripture to accommodate new discoveries. Still others sought a middle ground, attempting to synthesize revealed and natural truth.

From Dragons to Dinosaurs: A Transformed

Understanding

The Crystal Palace dinosaur sculptures of 1854 marked a watershed moment. Owen himself directed their construction, even hosting a famous New Year's Eve dinner inside the hollow Iguanodon model. These concrete creatures transformed dinosaurs from academic curiosities into cultural phenomena.

But something was lost in translation. The biblical "dragons" (tanninim in Hebrew) that had seemed mysterious but comprehensible became prehistoric monsters from an alien world. Where previous generations might have seen Leviathan or Behemoth in fossil bones, Victorians saw evidence of a pre-human earth that challenged traditional readings of Scripture.

Some Christians began searching Scripture for dinosaur references. Could Behemoth in Job 40:15-24 describe a sauropod dinosaur? "Look at Behemoth, which I made along with you... His tail sways like a cedar." Could Leviathan in Job 41 be a marine reptile? And what about the dragons (tanninim) mentioned throughout the Old Testament—were these dinosaurs known to ancient peoples?

These interpretations remain debated. Critics note that ancient Hebrew had limited vocabulary for exotic animals, often using "dragon" for any large, dangerous creature. Supporters argue that some descriptions fit dinosaurs better than any living animal.

By century's end, three broad camps had emerged. The fundamentalist reaction rejected geological timescales and maintained young earth and recent creation. Liberal accommodation fully accepted science and treated Genesis as allegory or myth. Those seeking middle ground looked for interpretations honoring both Scripture and science.

Each group claimed faithfulness to truth, but they defined truth differently. The stage was set for conflicts that continue today.

Lessons from History

The early Christian paleontologists offer both inspiration and caution. Their intellectual courage in pursuing truth wherever it led, their reverent wonder at God's glory in creation's vastness, and their faithful scholarship integrating faith and learning provide positive examples we should emulate.

Yet we should also heed the warnings their stories provide. Some harmonizations created more problems than they solved through hasty conclusions. Defensiveness often damaged Christian credibility when fear-based reactions took hold. And forcing false dichotomies between faith and science hurt both.

From these pioneers, we can distill timeless wisdom. First, the God of truth does not fear truth—wherever it's found. Second, we must distinguish facts from interpretations—bones don't speak; people do. Third, unity matters more than unanimity—Christians can disagree charitably. Finally, the Creator transcends our categories—His methods may surprise us.

The first Christian paleontologists opened a door to deep time and alien worlds. They couldn't have imagined where their discoveries would lead—from simple "terrible lizards" to feathered dinosaurs, from static fossils to dynamic ecosystems, from three known species to over a thousand.

What would Mary Anning think, knowing her "sea dragon" launched such debates? Perhaps she'd remind us of Job's wisdom: creation itself teaches those willing to learn. The rocks cry out, as Jesus said—sometimes literally, in the form of fossils.

The question isn't whether Christians should engage with dinosaur discoveries. The question is how we engage: with fear or faith, with dogmatism or humility, with conflict or wisdom?

Looking Ahead

As we prepare to explore what modern science tells us about dinosaurs, remember these pioneers. They studied ancient bones not despite their faith, but because of it. They found rooms in God's house they never expected, filled with creatures beyond imagination.

In the next chapter, we'll examine what we actually know about dinosaurs today. How do scientists study creatures dead for millions of years? What can fossils really tell us? And how do we separate scientific facts from speculative theories?

The journey from Buckland's Megalosaurus to today's discoveries is remarkable. But even more remarkable is how these ancient creatures continue to challenge and inspire our understanding of the Creator who made both them and us.

For Reflection

1. How do you think you would have reacted to dinosaur discoveries in the 1800s?
2. What can we learn from Christians who saw no conflict between their faith and their scientific work?
3. How might humility help us navigate questions about dinosaurs and faith today?

Further Reading

* Cadbury, Deborah. The Dinosaur Hunters
* Rudwick, Martin. Worlds Before Adam
* Numbers, Ronald. The Creationists (expanded edition)

Did You Know?

The word "paleontology" comes from Greek words meaning "ancient" (palaios), "being" (on), and "study" (logos). It literally means "the study of ancient life"–a discipline largely founded by Christians seeking to understand God's creative work across deep time.

Historical Note

The first dinosaur models at Crystal Palace (1854) depicted the creatures as gigantic lizards walking on four legs. Richard Owen even hosted a dinner party inside the hollow Iguanodon model! Our understanding has changed dramatically–we now know Iguanodon walked primarily on two legs–but the sense of wonder remains.

Chapter 2:

What We Really Know About Dinosaurs

"Great are the works of the Lord; they are pondered by all who delight in them." — Psalm 111:2

The Bone Hunters' Art

Dr. Mary Schweitzer stared at the microscope image in disbelief. It was 2005, and she was examining a 68-million-year-old Tyrannosaurus rex femur from Montana. But what she saw shouldn't exist: soft, flexible tissue and what appeared to be blood vessels. After 68 million years, only hard minerals should remain. Yet here was organic material, defying everything paleontologists thought they knew about fossilization.

"I looked at it and I said, 'This can't be,'" Schweitzer later recalled. She spent weeks trying to prove herself wrong. But the evidence held: somehow, soft tissue had survived in deep time.

This discovery would revolutionize paleontology and reignite debates about the age of dinosaurs. It perfectly illustrates both what we know and what we don't know about these ancient creatures. Science progresses not through dogma but through evidence—even when that evidence challenges established theories.

So what do we really know about dinosaurs? How do scientists study creatures that lived so long ago? And perhaps most importantly for Christians seeking truth: how reliable are these methods?

How Paleontologists Study the Past

Imagine trying to understand a civilization using only broken pottery, scattered bones, and faint impressions in rock. That's the paleontologist's challenge. Every fossil is a clue in a vast mystery story, and scientists have developed ingenious methods to extract information from these ancient remains.

When paleontologists discover dinosaur fossils, they don't simply dig them up and cart them away. The process is meticulous. Every bone's position is mapped using GPS and 3D scanning. The surrounding rock layers tell crucial stories about environment and age, so context recording becomes essential. Fossils are often jacketed in plaster for protection during the careful extraction process. Then months or years of laboratory preparation reveal hidden details.

Consider the famous "Fighting Dinosaurs" specimen from Mongolia—a Velociraptor and Protoceratops locked in mortal combat, preserved by a sudden sandstorm. The positioning of every claw and tooth tells a story frozen in time. Without careful excavation, this narrative would be lost.

Fossils don't come with labels saying "75 million years old" or "lived in swamps." Scientists deduce this information through multiple lines of evidence. Rock layers form in sequence, with older layers beneath younger ones. By studying these layers worldwide, geologists have constructed a "geological column" showing the relative ages of different fossils.

Environmental clues abound in the rocks. Ripple marks indicate ancient streams, while mud cracks suggest drying pools. Fossil plants reveal climate conditions, and associated animals indicate ecosystems. For instance, when Spinosaurus fossils are found with ancient crocodiles and lungfish, scientists deduce it lived in river systems. The shape of its teeth—conical, like a crocodile's—confirms a fish-eating diet.

Today's paleontologists use tools unimaginable to Buckland or

Mantell. CT scanning reveals internal structures without damaging fossils. Scientists discovered that Tyrannosaurus rex had an excellent sense of smell by CT scanning its braincase. Isotope analysis examines different isotopes in bones and teeth to reveal diet and climate. Analysis of sauropod teeth showed some species migrated hundreds of miles seasonally. Computer modeling allows simulations to test how dinosaurs moved. T. rex couldn't run faster than 25 mph without breaking its legs—bad news for Jurassic Park, good news for time travelers! And molecular paleontology, opened by Schweitzer's soft tissue discovery, has found proteins, pigment cells, and even possible DNA fragments in dinosaur fossils.

Dating Methods: Promises and Problems

Perhaps no aspect of paleontology generates more controversy among Christians than dating methods. How do scientists determine that dinosaurs lived millions of years ago? What are the assumptions? What are the limitations?

Radiometric dating works on a simple principle: radioactive elements decay at predictable rates. By measuring the ratio of parent to daughter elements, scientists calculate how long decay has been occurring. The discovery of radioactivity in 1896 revolutionized geology, providing potential "clocks" for dating rocks.

Different radioactive elements decay at different rates, measured in half-lives—the time required for half the parent isotope to decay. Carbon-14 has a half-life of 5,730 years, useful for recent materials. Potassium-40's half-life of 1.25 billion years dates older rocks. Uranium-238, with a 4.5-billion-year half-life, can date Earth's oldest minerals.

For dinosaur fossils, several methods apply. The bones themselves can't be directly dated—fossilization replaces original material with minerals. Instead, scientists date volcanic ash layers above and below fossil-bearing sediments. If ash below dinosaur bones dates to 70 million years and ash above dates to 65 million years, the dinosaurs lived between those times.

Multiple dating methods often cross-check each other. The K-T boundary (now called K-Pg boundary) marking dinosaur extinction has been dated by potassium-argon, argon-argon, uranium-lead, and rubidium-strontium methods at numerous sites worldwide. All converge on approximately 66 million years ago. This consistency across methods and locations builds confidence in the results.

However, radiometric dating requires several assumptions that young-earth creationists challenge. First, we must know the starting ratio of parent to daughter elements—the initial conditions. Conventional geologists argue that daughter products can be identified and excluded, but critics question whether we can truly know conditions millions of years ago. Contamination or inheritance of daughter isotopes would yield falsely old ages.

Second, the rock must remain closed to addition or removal of parent or daughter isotopes except by radioactive decay—the closed system assumption. But rocks experience weathering, metamorphism, and groundwater infiltration. Critics argue these processes could alter isotope ratios, making ages unreliable. Defenders counter that careful sample selection and multiple analyses can detect and avoid altered rocks.

Perhaps most controversially, radiometric dating assumes decay rates never change. This seems reasonable—extensive testing shows decay rates unaffected by temperature, pressure, or chemical environment. But young-earth creationists propose past conditions might have accelerated decay dramatically.

The RATE project (Radioisotopes and the Age of The Earth), conducted by creation scientists, investigated whether decay rates might have been millions of times faster in the past. They point to several lines of evidence: helium retention in zircons suggesting accelerated alpha decay, radiohalos indicating rapid uranium decay, and discordant isochron dates possibly reflecting non-constant decay.

The Young-Earth Challenge

Some Christians point to anomalies:

- C-14 found in dinosaur bones (shouldn't exist if they're millions of years old)
- Soft tissue preservation (how could it last so long?)
- Rapid formation of some rock layers

These deserve serious consideration. Possible explanations include:

- Contamination by modern carbon
- Unknown preservation mechanisms
- Catastrophic local events within long timescales

The debate continues, with sincere scientists on multiple sides seeking truth.

The Spectacular Diversity of Dinosaurs

When most people think "dinosaur," they picture T. rex or Brachiosaurus. But dinosaurs were as diverse as modern mammals, ranging from crow-sized Microraptor to 100-foot Argentinosaurus. The smallest, Parvicursor, weighed less than a pound, while the largest, Argentinosaurus, could reach 100 tons. Most dinosaurs had body sizes between a chicken and an elephant.

Their dietary diversity was equally impressive. Carnivores like Allosaurus possessed blade-like teeth for slicing flesh. Herbivores such as Triceratops had grinding batteries of hundreds of teeth. Omnivores like Ornithomimus showed varied dental patterns. Even filter feeders existed—Pterodaustro, technically a pterosaur, had sieve-like mouths for straining small organisms from water.

Dinosaurs adapted to every available habitat. Protoceratops thrived in desert environments. Cryolophosaurus lived in polar regions that, while warmer than today's Antarctica, still experienced months of darkness. Island environments produced dwarf species like Magyarosaurus, demonstrating the same evolutionary pressures that create island dwarfism today. Various Chinese discoveries show mountain-dwelling species adapted to high altitudes.

Modern techniques reveal surprising behaviors. Maiasaura ("good mother lizard") fossils show nest colonies resembling those of modern seabirds, with adults bringing food to young and extended parental care periods. Track sites reveal fascinating social structures—sauropod herds traveled with young protected in the center, Deinonychus hunted in coordinated packs, and Tyrannosaurus may have formed family groups.

Dinosaurs also developed complex communication methods. Parasaurolophus's elaborate crest functioned as a resonating chamber for producing sounds. Styracosaurus's impressive horn array served for visual display and species recognition. Recent discoveries even suggest some species had color patterns preserved in fossilized melanosomes, indicating visual communication was as important then as now.

Myths Versus Facts

Popular misconceptions about dinosaurs abound. Contrary to Hollywood depictions, most dinosaurs were smaller than a car, not the giants we imagine. Far from being slow, stupid reptiles, many were active creatures with complex behaviors, possibly warm-blooded.

The timeline of dinosaur existence surprises many. All dinosaurs didn't live at the same time—Stegosaurus was already fossilized when T. rex appeared. More time separates Stegosaurus from

T. rex than separates T. rex from us! Rather than evolutionary failures, dinosaurs dominated Earth for 165 million years, while humans have existed for merely 0.3 million years. And not all dinosaurs went extinct—birds are living dinosaurs, as we'll explore in Chapter 11.

Movies shape public perception, but entertainment trumps accuracy. The Velociraptors in Jurassic Park stood 6 feet tall with scaly skin, while real Velociraptors were 3 feet tall and fully feathered. The movie's notion that T. rex vision was based on movement is pure fiction—fossil evidence indicates excellent binocular vision. And those terrifying roars and shrieks? Real dinosaurs probably made sounds more like modern birds—chirps, hisses, and booming calls.

Revolutionary Recent Discoveries

The discovery of feathered dinosaurs in China since the 1990s revolutionized paleontology. Sinosauropteryx, discovered in 1996, was the first non-bird dinosaur with confirmed feathers—simple, hair-like proto-feathers with preserved pigments showing reddish-brown stripes. Microraptor, found in 2003, possessed four wings with feathers on both arms and legs, colored iridescent black like modern ravens, and capable of gliding between trees. Even large dinosaurs bore feathers—Yutyrannus, discovered in 2012, was a 30-foot feathered tyrannosaur proving that size didn't preclude plumage, living in cooler climates where insulation was beneficial.

These discoveries blur the line between dinosaurs and birds, raising questions about classification and evolution.

Scientists can now determine dinosaur colors by analyzing melanosomes—pigment-containing structures in fossilized feathers. Anchiornis had a black body with a reddish head crest. Sinosauropteryx displayed reddish-brown coloring with a striped tail. Even the armored Psittacosaurus showed complex camouflage patterns. This isn't speculation—it's based on comparing fossilized structures to those in modern bird feathers.

Mary Schweitzer's soft tissue discoveries opened new frontiers and controversies. Scientists have found flexible blood vessels, possible red blood cells, proteins including collagen, and even possible DNA fragments. The controversy centers on how organic material could survive millions of years. Does this evidence support a young Earth? Current scientific thinking suggests iron from blood may act as a preservative, exceptional conditions can preserve organics far longer than expected, but more research is needed on preservation mechanisms.

What We Know, What We Don't, What We May Never Know

Some dinosaur facts are well-established through multiple lines of evidence. We know they existed—thousands of specimens from every continent prove this beyond doubt. Their general time periods show consistent dating from multiple methods. Basic anatomy is clear from skeletal fossils, and some behaviors are evidenced by tracks, nests, and bite marks. The diversity is staggering, with over 1,000 valid species identified and counting.

Yet scientists actively debate many aspects. Were dinosaurs warm-blooded, cold-blooded, or something unique? How fast did they grow to reach adult size? What caused their extinction—asteroid impact, volcanism, or a combination of factors? When and why did feathers first appear? How sophisticated were dinosaur herds and packs in their social behaviors?

Some questions may remain forever unanswered. Soft tissue rarely preserves, so we may never know what sounds dinosaurs made. Behaviors like mating displays leave few fossil traces. Brain size doesn't tell us everything about intelligence levels. Diseases and parasites are usually invisible in fossils. And except for some feathered species, we can only guess at exact colors and patterns.

Implications for Faith

What does this mean for Christians seeking to understand God's creation?

The history of paleontology teaches humility. Scientific understanding constantly evolves—dinosaurs were once thought to be lizards, then they became active animals, and now many are known to have feathers. Tomorrow may bring new surprises. This doesn't mean science is unreliable—it means science is self-correcting. Good science, like good theology, requires humility.

Every new discovery reveals greater complexity. We find sophisticated ecosystems, intricate anatomical designs, complex behaviors, and unexpected preservation. Whether dinosaurs lived thousands or millions of years ago, they display the Creator's incredible creativity.

We know more about dinosaurs than ever, yet mysteries multiply. Each answer raises new questions. This isn't a weakness—it's the nature of studying God's vast creation. As Proverbs 25:2 says, "It is the glory of God to conceal a matter; to search out a matter is the glory of kings."

Looking Forward

We've surveyed what science tells us about dinosaurs—the methods, the discoveries, the debates. But how do these ancient creatures fit into the Biblical narrative? Did the writers of Scripture know about dinosaurs? Could Behemoth and Leviathan be dinosaurian?

In the next chapter, we'll explore "Dragons in the Bible and Ancient Cultures," examining texts and traditions that some believe preserve memories of dinosaurs. The intersection of paleontology and Scripture raises fascinating questions that deserve careful, respectful consideration.

Whether dinosaurs died 66 million years ago or more recently, whether dating methods are reliable or flawed, one truth remains: these creatures existed, they were magnificent, and they declare the glory of their Creator. In seeking to understand them, we join the psalmist in pondering the works of the Lord.

For Reflection

1. How does learning about paleontological methods affect your view of dinosaur science?
2. What aspects of dinosaur diversity most reveal God's creativity to you?
3. How can Christians maintain humility while seeking truth about Earth's history?

Did You Know?

The largest dinosaur eggs ever found are only about the size of a football. Even 100-ton sauropods started life smaller than a house cat! This limitation (eggs can only be so big before shells become too thick) may have influenced dinosaur reproduction strategies.

Incredible Fact

Dracorex hogwartsia (meaning "Dragon King of Hogwarts") is a real dinosaur named in 2006. The skull looked so dragon-like that scientists gave it a name honoring Harry Potter! Some paleontologists think it might be a juvenile Pachycephalosaurus rather than a separate species–showing how science continues to refine our understanding.

How We Know

When scientists say T. rex had excellent smell, they're not guessing. CT scans of skull cavities show huge olfactory bulbs–the brain regions processing smell. By comparing the ratio to modern animals, paleontologists can estimate sensory capabilities. T. rex's nose was as keen as a modern vulture's!

For Further Study

* Fastovsky & Weishampel: Dinosaurs: A Concise Natural History
* Schweitzer's papers on soft tissue preservation
* Online: American Museum of Natural History's paleontology resources
* Documentary: "Dinosaur 13" (the story of "Sue" the T. rex)

Chapter 3:

Dragons in the Bible and Ancient Cultures

"Speak to the earth, and it will teach you; let the fish of the sea declare to you. Who among all these does not know that the hand of the Lord has done this?" — Job 12:8-9

The Petroglyph Mystery

In the remote canyons of Natural Bridges National Monument in Utah, archaeologist Samuel Hubbard made a startling discovery in 1924. Among the ancient Ancestral Puebloan petroglyphs was an image that seemed impossible: a creature with a long neck, thick body, and long tail that looked remarkably like a sauropod dinosaur. How could people who lived perhaps a thousand years ago depict an animal that science says went extinct 66 million years before humans appeared?

Similar mysteries appear worldwide. The Angkor Wat temple in Cambodia, built in the 12th century, features a stone carving that some interpret as a Stegosaurus. Ancient Mesopotamian cylinders depict creatures resembling long-necked dragons. Aboriginal Australian artwork shows creatures that some compare to Megalania, a giant monitor lizard thought extinct for 40,000 years.

These artifacts raise profound questions. Are they evidence that humans and dinosaurs coexisted? Do they represent cultural memories passed down from prehistoric encounters? Or are we seeing patterns where none exist, like finding animals in clouds?

For Christians seeking to understand dinosaurs in light of Scripture, these questions become even more significant when we turn to the Bible itself. The Hebrew Scriptures speak of mysterious creatures—tanninim, Behemoth, Leviathan—that defy easy identification. Could the Bible be describing dinosaurs?

Dragons in Scripture: Lost in Translation

The Hebrew Word "Tanninim"

Throughout the Old Testament, we encounter the Hebrew word "tanninim" (plural of "tannin"), variously translated as dragons, serpents, sea monsters, or jackals. The word appears in contexts ranging from creation to prophecy, and its meaning shifts like sand depending on the translator's interpretation.

In Genesis 1:21, God creates the "great tanninim" on the fifth day. The King James Version boldly calls them "great whales," while modern translations prefer "great sea creatures" or "great sea monsters." The Hebrew text offers no clarification—these are simply the great tanninim of the waters. Were they whales? Prehistoric marine reptiles? Something else entirely? The text preserves the mystery.

The word appears again when Moses confronts Pharaoh. In Exodus 7:9-12, both Moses' staff and the Egyptian magicians' staffs become tanninim—usually translated as "serpents" here, though it's the same Hebrew word used for the sea creatures in Genesis. This flexibility in meaning suggests tannin referred to various impressive reptilian creatures rather than a specific species.

Perhaps most intriguingly, tanninim appear in poetic passages describing God's power over chaos. Psalm 74:13 declares, "You divided the sea by your might; you broke the heads of the tanninim on the waters." Isaiah 27:1 prophesies that God will punish "Leviathan the fleeing serpent, Leviathan the twisting serpent, and he will slay the tannin that is in the sea."

These passages blend natural history with cosmic symbolism. Ancient Near Eastern cultures often depicted chaos as a dragon or sea monster that the gods must defeat. The biblical writers adopted this imagery while insisting that Yahweh, not the pagan gods, controls these creatures. But were they purely symbolic, or did they reference actual animals known to ancient peoples?

The Behemoth of Job 40

The book of Job contains the Bible's most detailed descriptions of mysterious creatures. After Job questions God's justice, the Lord responds with a tour of creation's wonders, culminating in two spectacular beasts: Behemoth and Leviathan. These passages have sparked centuries of debate and, more recently, speculation about dinosaurs.

God introduces Behemoth in Job 40:15-24 with remarkable fanfare: "Behold, Behemoth, which I made as I made you; he eats grass like an ox." This opening emphasizes that Behemoth is a created animal, not a mythological being. It shares the earth with Job, eating vegetation like familiar livestock.

But the description quickly moves beyond the ordinary. "Behold, his strength in his loins, and his power in the muscles of his belly. He makes his tail stiff like a cedar; the sinews of his thighs are knit together." Here controversy erupts. The comparison of Behemoth's tail to a cedar tree has become a battleground in the dinosaur debate.

Traditional interpreters often identify Behemoth as a hippopotamus or elephant. These were the most powerful land animals known to ancient readers. The description of strength, the river habitat mentioned later in the passage, and the herbivorous diet all fit. But critics point out that neither hippos nor elephants have cedar-like tails—their tails are relatively small and rope-like.

Could Behemoth be a sauropod dinosaur? Advocates note that many sauropods had massive, muscular tails that could indeed resemble cedar logs. Diplodocus used its tail like a whip; Shunosaurus even had a tail club. The description of bones "like tubes of bronze" and limbs "like bars of iron" certainly fits the massive leg bones of sauropods.

The passage continues with vivid details. "For the mountains yield food for him where all the wild beasts play. Under the lotus plants he lies, in the shelter of the reeds and in the marsh." This creates an almost Jurassic scene—a massive herbivore feeding while other animals play nearby, resting in marshy areas under prehistoric-looking plants.

Yet we must be cautious. Hebrew poetry often uses hyperbole and metaphor. When the text says "He is the first of the works of God," it may mean the most impressive rather than literally the first created. When it claims "let him who made him bring near his sword," suggesting only God can threaten Behemoth, this poetic language emphasizes power rather than providing zoological precision.

Leviathan: The Untameable Dragon

If Behemoth impresses with size and strength, Leviathan terrifies with ferocity and power. Job 41 dedicates an entire chapter to this creature, the longest animal description in Scripture. Unlike Behemoth's brief portrait, Leviathan receives an epic treatment that has captivated readers for millennia.

The description begins with a series of rhetorical questions emphasizing human powerlessness. "Can you draw out Leviathan with a fishhook or press down his tongue with a cord? Can you put a rope in his nose or pierce his jaw with a hook?" The ancient practice of catching and controlling large aquatic animals—from fish to crocodiles—becomes laughable when applied to Leviathan.

As the description unfolds, Leviathan's features become increasingly spectacular. His hide is impenetrable armor: "Who can penetrate his double coat of mail? Who can open the doors of his face? Around his teeth is terror." Weapons prove useless: "The arrow cannot make him flee; for him sling stones are turned to stubble. Clubs are counted as stubble; he laughs at the rattle of javelins."

Most dramatically, Leviathan breathes fire: "Out of his mouth go flaming torches; sparks of fire leap forth. Out of his nostrils comes forth smoke, as from a boiling pot and burning rushes. His breath kindles coals, and a flame comes forth from his mouth." This detail has provoked endless debate. Is this pure poetry, or could it describe a real phenomenon?

Traditional identifications include the crocodile, which fits the aquatic habitat and fearsome reputation. Ancient Egyptians both worshipped and feared Nile crocodiles. But crocodiles don't breathe fire or smoke. Some suggest poetic exaggeration of the steam that rises from a crocodile's nostrils in cool morning air, but this seems a stretch.

Those who see dinosaurs in Job propose various extinct reptiles. Marine reptiles like Kronosaurus or Mosasaurus were apex predators of ancient seas. The description of Leviathan "making the deep boil like a pot" and leaving "a shining wake behind him" could describe the turbulence created by a massive marine predator.

The fire-breathing aspect remains problematic for any naturalistic interpretation. However, some creation scientists note that the bombardier beetle produces a boiling chemical spray for defense. They speculate that extinct reptiles might have possessed similar but more dramatic chemical defense mechanisms. While intriguing, no fossil evidence supports fire-breathing capabilities in any known creature.

Perhaps most significant is Leviathan's symbolic role. Throughout Scripture, Leviathan represents chaos and evil that only God can control. Psalm 74:14 speaks of God crushing "the heads of Leviathan"—note the plural, suggesting a many-headed dragon. Isaiah uses Leviathan as a symbol for enemy nations God will judge. By Job 41, this chaos-dragon has become a created animal under God's sovereignty, powerful but not divine.

Ancient Dragons Around the World

Universal Dragon Myths

The prevalence of dragon legends across cultures presents an anthropological puzzle. From Chinese lung to European wyrms, from Aztec Quetzalcoatl to Aboriginal Rainbow Serpents, nearly every ancient culture features serpentine monsters. These creatures share remarkable similarities despite arising in isolated societies: they're typically reptilian, often enormous, sometimes flying, occasionally breathing fire or poison, and frequently associated with water.

Conventional anthropology explains this convergence through common psychological factors. Humans instinctively fear snakes, crocodiles, and large predators. Combining these fears creates the ultimate monster—the dragon. Storm clouds writhing across the sky might inspire legends of celestial serpents. Finding large fossil bones without understanding their origin could generate tales of giants and dragons.

But some researchers propose a more direct explanation: dragon legends preserve actual encounters with prehistoric survivors. They note specific details in some legends that seem too accurate for pure imagination. The Aboriginal stories of the "bunyip" describe a large, swamp-dwelling creature with a booming call—features that match the extinct marsupial Diprotodon. Chinese dragon imagery, with its four legs and long body, resembles certain dinosaur reconstructions more than any living reptile.

The Stones of Ica and Other Controversial Artifacts

Peru's Ica Stones present one of the most debated examples of possible human-dinosaur interaction. This collection of thousands of engraved stones depicts various scenes, including clear representations of dinosaurs interacting with humans. Some show men riding Triceratops-like creatures; others depict hunters attacking long-necked sauropods.

The stones' authenticity remains hotly contested. Skeptics note that local farmers admitted to creating fake stones for tourists, and the patina can be artificially aged. The dinosaur depictions suspiciously match mid-20th century reconstructions, complete with errors later corrected by paleontology. No verified archaeological context exists for the stones.

Supporters counter that some stones show anatomical details unknown when they surfaced, such as dermal spines on sauropods later confirmed by fossils. They argue that while fake stones certainly exist, this doesn't invalidate the entire collection. Similar controversies surround the Acambaro figurines from Mexico and various Native American petroglyphs.

The Angkor Wat stegosaur carving stands apart because its context is unquestionable—it's carved into a 12th-century temple. The image shows a quadrupedal creature with what appears to be plates along its back. Skeptics identify it as a chameleon or rhino viewed from above, with decorative elements misinterpreted as plates. Believers see a clear Stegosaurus depiction. The truth likely depends on the observer's preconceptions.

Dragons in Historical Records

Moving beyond prehistory, historical records contain intriguing accounts of dragon encounters. Medieval chronicles describe these not as supernatural beings but as dangerous animals requiring extermination. The Anglo-Saxon Chronicle matter-of-factly records dragon sightings alongside battles and coronations.

St. George's dragon, while embellished by legend, may have historical roots. Early versions describe a large reptile terrorizing a Libyan town, not the winged, fire-breathing monster of later art. Similarly, Beowulf's dragon is described more like a large serpent guarding treasure in an earthen mound than a fantastical beast.

Marco Polo's travels include descriptions of "huge serpents" in China with "two short legs with three claws," jaws "large enough to swallow a man in one gulp," and "very savage" temperaments. While possibly describing crocodiles, some details don't match any known species.

These accounts face the challenge of medieval credulity. The same chronicles that record dragons also describe unicorns, basilisks, and other mythical creatures. Yet the matter-of-fact tone used for dragon encounters, often including specific locations and dates, differs from the obvious mythology surrounding other fantastic beasts.

Interpreting the Evidence

The Case for Dinosaurs in Human History

Those who believe humans encountered dinosaurs marshal several lines of evidence. The universal nature of dragon legends suggests a common source beyond imagination. Specific anatomical details in ancient art sometimes match fossil discoveries made centuries later. The biblical descriptions of Behemoth and Leviathan contain details that fit extinct reptiles better than any living animals.

They argue that the conventional timeline must be wrong. If humans and dinosaurs coexisted, whether in recent history or the distant past, it would explain these cultural memories. Some propose that small populations of dinosaurs survived the mass extinction, persisting in remote areas into human history. Others maintain that all history fits within a young-earth timeline, making coexistence inevitable.

This view faces significant challenges. No verified dinosaur remains have been found in archaeological contexts with human artifacts. The fossil record shows a clear separation between dinosaur-bearing and human-bearing rock layers. Dragon legends, while widespread, vary enormously in details and often include clearly mythological elements.

Alternative Explanations

Mainstream scholarship offers various explanations for dragon legends without requiring human-dinosaur coexistence. The discovery of fossil bones by ancient peoples could inspire monster myths. A Protoceratops skull, with its beak and neck frill, might suggest a griffin. Mammoth skulls, with their central nasal cavity, possibly inspired cyclops legends.

Living animals provide ample inspiration for dragons. Crocodiles, monitor lizards, and large snakes combine reptilian features with genuine danger. The extinct Megalania, a 20-foot monitor lizard, survived until humans reached Australia. Such encounters could generate lasting cultural memories exaggerated over generations.

Psychological factors also contribute. Humans possess innate fears of specific dangers—snakes, big cats, birds of prey—that protected our ancestors. Dragons embody all these fears in one ultimate predator. This explains why independent cultures created similar monsters.

The biblical creatures might represent known animals described in poetic language. Behemoth could be a hippopotamus with hyperbolic flourishes. Leviathan might combine crocodile features with mythological chaos-monster imagery common in ancient Near Eastern literature.

A Thoughtful Christian Approach

For Christians seeking truth, this topic requires careful balance. We must avoid two extremes: dismissing potentially significant evidence because it challenges conventional timelines, or accepting every dragon legend as proof of recent dinosaurs.

Scripture's authority doesn't depend on Behemoth being a dinosaur. The Job passages primarily teach God's sovereignty over creation, not natural history. Whether Behemoth represents a hippo, a sauropod, or a poetic composite, the theological message remains unchanged: God created awesome creatures that humble human pride.

Similarly, dragon legends prove human creativity and cultural exchange as much as any historical encounters. Even if some legends preserve genuine memories of extinct megafauna, this wouldn't necessarily validate every dragon story or artifact.

The key is maintaining intellectual honesty while remaining open to evidence. If humans did encounter dinosaur survivors, it would revise our understanding of natural history without affecting core Christian doctrine. If dragon legends arose from fossils, fear, and imagination, God remains sovereign over both the terrible lizards of the Mesozoic and the human minds that reimagined them as dragons.

Lessons from the Dragon Debate

The question of dragons and dinosaurs teaches valuable lessons about interpreting both Scripture and science. First, we must respect the literary genres of biblical texts. Job's poetic descriptions differ from historical narrative. Expecting zoological precision from Hebrew poetry misunderstands the text's purpose.

Second, we should evaluate evidence carefully without predetermined conclusions. Some dragon artifacts are clearly fake; others remain genuinely puzzling. Each deserves individual assessment rather than wholesale acceptance or rejection.

Third, this debate reminds us that much about the past remains mysterious. Whether concerning dinosaur extinction, ancient human cultures, or the intersection of natural and supernatural history, humility serves us better than dogmatism.

Finally, these questions, while fascinating, shouldn't divide Christians. Whether Leviathan was a crocodile or a Kronosaurus, whether dragons preserve dinosaur memories or mythological archetypes, the central truths of faith remain unaffected. God created all things, judges righteously, and offers salvation through Christ—truths more solid than any fossilized bone or ancient carving.

Looking Ahead

We've explored the tantalizing possibility that biblical and cultural dragons might preserve memories of dinosaurs. The evidence remains ambiguous, open to multiple interpretations. In our next chapter, we'll examine how young-earth creationists build their case for human-dinosaur coexistence within a biblical timeline. Their model, while controversial, deserves careful consideration as one Christian approach to reconciling Scripture with the terrible lizards that once ruled the earth.

For Reflection

1. How do you interpret the descriptions of Behemoth and Leviathan in Job? What criteria guide your interpretation?
2. Why do you think dragon legends appear in so many cultures? What does this tell us about human nature and cultural development?
3. How can Christians evaluate controversial evidence like the Ica Stones while maintaining both skepticism and openness?

Did You Know?

The bombardier beetle mixes chemicals in its abdomen to produce a boiling hot spray for defense. The reaction reaches 212°F (100°C) and can be aimed precisely at attackers. While not "breathing fire," this mechanism shows that biological chemical weapons exist in nature–though no fossil evidence suggests any reptile possessed such abilities.

Food for Thought

The Hebrew word "rahab" appears in Scripture as both a sea monster associated with chaos (Job 9:13, Psalm 89:10) and as a symbolic name for Egypt (Psalm 87:4, Isaiah 30:7). This dual usage–mythological and geopolitical–reminds us that ancient writers freely mixed literal and symbolic language in ways foreign to modern categories.

Further Reading

* Mayor, Adrienne. The First Fossil Hunters
* Lyons, Eric. Behemoth and Leviathan–Creatures of Controversy
* Isaacs, David. Dragons or Dinosaurs?: Creation or Evolution?
* Shuker, Karl. Dragons: A Natural History

Chapter 4:

"Dinosaurs Lived with Adam": The Young-Earth View

"For in six days the Lord made heaven and earth, the sea, and all that is in them, and rested on the seventh day." — Exodus 20:11

A Museum Unlike Any Other

The Creation Museum in Petersburg, Kentucky, draws nearly half a million visitors annually to see an exhibition that would have astounded both Charles Darwin and Richard Owen. Here, animatronic dinosaurs share garden paths with figures of Adam and Eve. A Triceratops with a saddle suggests humans domesticated dinosaurs. Display cases explain how T. rex used its fearsome teeth to crack open coconuts in Eden.

For many Christians, such scenes seem like fantasy—a well-meaning but misguided attempt to reconcile faith with fossils. For others, they represent essential truth: God created dinosaurs alongside humans during the creation week, just 6,000 years ago. This young-earth creationist view, while minority in the scientific community, claims millions of adherents worldwide and deserves thoughtful examination.

The movement's modern form began with The Genesis Flood (1961) by John Whitcomb and Henry Morris. This influential book argued that proper biblical interpretation requires a recent creation and global flood. Dinosaurs, therefore, must fit within this framework—created on Day Six, surviving on Noah's ark, and living into human history before eventual extinction.

To understand this perspective fairly, we must examine its biblical foundations, scientific arguments, and responses to challenges. Whether one agrees or disagrees, the young-earth model represents a serious attempt to maintain biblical authority while explaining the fossil record.

The Biblical Foundation

Reading Genesis Literally

Young-earth creationists begin with what they see as the plain meaning of Scripture. Genesis 1 describes God creating the universe in six days, with evening and morning marking each day. The Hebrew word "yom" (day), when modified by a number and the phrase "evening and morning," always means a literal 24-hour day elsewhere in Scripture.

Exodus 20:11 seems to confirm this interpretation. In giving the Sabbath commandment, God says He made everything "in six days" and rested the seventh. If creation days were actually millions of years, the commandment loses coherence. How can humans rest one day after working six if God's "days" were eons?

The genealogies of Genesis 5 and 11 provide another crucial piece. These chapters trace lineages from Adam to Abraham with specific ages and years. Adding these numbers, along with other biblical chronologies, yields a date for creation around 4000-6000 BC. Archbishop James Ussher famously calculated 4004 BC, and while modern young-earth creationists allow some flexibility, all place creation within the last 10,000 years.

This chronology leaves no room for dinosaurs living millions of years before humans. If the Bible is God's inerrant word, and if it clearly teaches recent creation, then human assumptions about deep time must yield to divine revelation. As Ken Ham, founder of Answers in Genesis, often says, "God's word is the starting point, not human opinions."

Death Before Sin

Perhaps the strongest theological argument for young-earth creationism concerns death's origin. Romans 5:12 states, "Therefore, just as sin came into the world through one man, and death through sin, and so death spread to all men because all sinned." If death entered through Adam's sin, how could millions of years of death and extinction precede humanity?

The fossil record documents massive carnage—predation, disease, cancer, and extinction. Dinosaur fossils show evidence of violence: Velociraptor locked in combat with Protoceratops, Tyrannosaurus bite marks on Triceratops bones, and mass death assemblages. How could God call such a world "very good" (Genesis 1:31)?

Young-earth creationists insist that the original creation was deathless, at least for sentient creatures. Animals with the "breath of life" (Hebrew: nephesh chayyah) couldn't die before sin. This category likely includes dinosaurs, making their fossils post-Fall phenomena. The apparent cruelty in the fossil record reflects sin's corruption, not God's original design.

The Global Flood as Key

If dinosaurs lived recently, why do we find their fossils in rock layers below human remains? The answer, according to young-earth geology, is Noah's flood. This catastrophic event provides both the mechanism for fossilization and explanation for fossil distribution.

Genesis 6-9 describes a global deluge that destroyed all land animals except those on the ark. The "fountains of the great deep" breaking up suggests massive tectonic activity. Torrential rain lasting 40 days and flood waters covering the earth for over a year would create ideal conditions for rapid burial and fossilization.

The order of fossil layers reflects not evolutionary progression but flood dynamics. Marine creatures were buried first, followed by lowland animals, with highland creatures last. Dinosaurs, being large and often living in lowland areas, were buried in middle layers. Humans, more mobile and intelligent, avoided burial until later stages or post-flood local catastrophes.

This model explains features that challenge conventional geology: polystrate fossils (trees extending through multiple rock layers), soft sediment deformation (showing rapid deposition), and mass fossil graveyards (requiring catastrophic burial). The flood provides a single, recent explanation for phenomena that standard geology attributes to millions of years.

Dinosaurs in Eden

Created on Day Six

According to the young-earth model, God created dinosaurs on the sixth day alongside other land animals and humans. The biblical text doesn't use the word "dinosaur"—Richard Owen wouldn't coin it for another 3,000+ years—but Genesis 1:24-25 describes God making "beasts of the earth according to their kinds."

The Hebrew word "tanniyn," often translated "sea monster" or "dragon," may refer to dinosaurs. When God creates "great sea creatures" (tanniyn) on Day Five, these could include marine reptiles like plesiosaurs and mosasaurs. The land dinosaurs would follow on Day Six as part of the "beasts of the earth."

In this view, Adam would have encountered dinosaurs in Eden. He named all the animals (Genesis 2:19-20), presumably including various dinosaur kinds. The original dinosaurs, like all creatures, were herbivorous. Genesis 1:30 states that God gave "every green plant for food" to all creatures with the breath of life.

This raises obvious questions about dinosaur anatomy. How did T. rex use teeth clearly designed for tearing flesh to eat plants? Young-earth creationists propose several possibilities. Perhaps carnivorous features were latent designs that only activated after the Fall. Or God redesigned creatures as part of the Curse. Some suggest that sharp teeth work well for tough plant material—pandas have sharp teeth but eat bamboo.

Living in Harmony

The pre-Fall world, in this model, was fundamentally different from today's ecology. No death, no predation, no scarcity meant radically different animal behavior. Lions played with lambs; Velociraptors grazed peacefully alongside Protoceratops. The competition and violence we see today results from sin's corruption, not original design.

This Edenic state lasted unknown duration—perhaps days, perhaps longer. Young-earth creationists don't require millions of years for geological processes but neither do they insist Adam sinned immediately. During this golden age, humans and dinosaurs coexisted in perfect harmony under God's direct providence.

The Fall changed everything. When Adam sinned, death entered creation. Thorns and thistles appeared (Genesis 3:18), suggesting fundamental biological changes. Animals began to fear humans (Genesis 9:2) and each other. The creation that was "very good" became groaning in bondage to decay (Romans 8:20-22).

The Ark and Dinosaur Survival

Fitting Dinosaurs on the Ark

Critics often ridicule the idea of dinosaurs on Noah's ark. How could Brachiosaurus or Argentinosaurus fit through the door? Where would Noah store food for hundreds of dinosaur species for a year? The logistics seem impossible.

Young-earth creationists have developed detailed responses. First, Noah didn't need to take every dinosaur species—only representative kinds. The biblical "kind" (Hebrew: min) likely corresponds roughly to the family level in modern taxonomy. Just as all dog breeds descended from a wolf-like ancestor, numerous dinosaur species could arise from fewer created kinds. This reduces the number dramatically.

Second, Noah would have taken juveniles, not adults. Even the largest dinosaurs began life smaller than a sheep. A juvenile Apatosaurus requires far less space and food than a full-grown adult. Reptiles also have indeterminate growth—they continue growing throughout life. Young specimens were ideal ark passengers.

Third, many dinosaurs were small. While we imagine giants, most dinosaur species were turkey-to-cow sized. The average dinosaur was about the size of a sheep. Popular media's focus on giants like T. rex distorts the actual diversity. The ark had ample room for representative juvenile pairs of all dinosaur kinds.

Some creationists propose that God might have induced hibernation or dormancy during the voyage. Many reptiles naturally undergo periods of reduced activity. Divine providence could have enhanced this tendency, minimizing food requirements and waste management challenges.

Post-Flood Survival and Extinction

According to this model, dinosaurs disembarked with Noah in the mountains of Ararat. The post-flood world, however, was radically different. Climate changes, reduced food sources, and human hunting pressure made survival difficult. While some dinosaur kinds persisted for centuries or millennia, most quickly went extinct.

The Ice Age, which young-earth geologists place immediately after the flood, created additional challenges. Warm oceans (from "fountains of the deep") and cool continents (from volcanic activity) generated intense precipitation and glaciation. Dinosaurs adapted to pre-flood conditions couldn't cope with rapid climate changes.

Human population growth added pressure. As people spread across the earth, they encountered dangerous dragons (dinosaurs) and hunted them. Medieval dragon slayers may have eliminated the last surviving dinosaurs in Europe. Similar patterns occurred worldwide, with humans driving dinosaurs to extinction just as they later did with mammoths and other megafauna.

Some young-earth creationists believe small populations might survive in remote areas. Reports of mokele-mbembe in Congo or lake monsters in various locations could represent relict dinosaurs. While mainstream science dismisses such cryptozoological claims, young-earth advocates remain open to the possibility.

Scientific Arguments for a Young Earth

Soft Tissue in Dinosaur Bones

Mary Schweitzer's discovery of soft tissue in T. rex fossils electrified the young-earth community. How could organic material survive 68 million years? Laboratory experiments show proteins degrade rapidly, even under ideal conditions. The presence of flexible blood vessels, apparent red blood cells, and intact proteins suggests these fossils are thousands, not millions, of years old.

Subsequent discoveries have found soft tissue in numerous dinosaur specimens. Collagen, elastin, laminin, and other proteins have been identified. DNA fragments may also be present. Each discovery strengthens the young-earth argument: these fossils cannot be as old as claimed.

Conventional scientists propose exotic preservation mechanisms—iron acting as a preservative, biofilms creating tissue-like structures, or unique geological conditions. Young-earth researchers find these explanations inadequate. The simpler solution is younger fossils. If dinosaurs lived recently, soft tissue preservation becomes expected rather than miraculous.

Carbon-14 in Dinosaur Bones

Perhaps even more significant is the reported discovery of carbon-14 in dinosaur fossils. C-14 has a half-life of 5,730 years, making it undetectable after about 100,000 years. Yet multiple laboratories have detected C-14 in dinosaur bones that should be C-14 dead if millions of years old.

The Paleochronology Group, composed of creation scientists, has funded C-14 testing of numerous dinosaur samples. Results consistently show detectable C-14, suggesting ages of thousands of years. When presented at scientific conferences, these findings provoke intense controversy.

Critics argue contamination explains these results. Modern carbon could infiltrate fossils through groundwater or handling. Young-earth researchers respond that they've taken extensive precautions against contamination. The consistency of results across multiple samples and laboratories suggests genuine endogenous carbon.

Rapid Geological Processes

Young-earth geologists argue that features attributed to millions of years can form rapidly under catastrophic conditions. Mount St. Helens' 1980 eruption provided a natural laboratory. In hours and days, the volcano produced stratified deposits, carved canyons, and created features resembling those supposedly requiring vast time.

Similar rapid processes are documented worldwide. Turbidity currents can deposit enormous sediment layers quickly. Volcanic ash can fossilize trees in weeks. Coal can form from plant matter in months under proper conditions. If local catastrophes produce such effects rapidly, a global flood would magnify these processes exponentially.

Polystrate fossils—usually trees extending through multiple rock layers—particularly challenge conventional geology. How did these trees remain upright for millions of years while sediment slowly accumulated? The young-earth answer is rapid deposition during the flood. Entire forests were buried quickly, preserving trees in growth position.

Answering Major Objections

Radiometric Dating Challenges

The most significant scientific challenge to young-earth creationism is radiometric dating. Multiple independent methods consistently yield ancient ages for dinosaur-bearing rocks. How can young-earth creationists dismiss such widespread evidence?

The RATE project (Radioisotopes and the Age of The Earth), conducted by creation scientists, examined this question intensively. They identified several key issues with radiometric dating assumptions. First, initial conditions are unknown. We cannot know the original parent-daughter isotope ratios in rocks. Different assumptions yield different ages.

Second, decay rates may not be constant. While this challenges fundamental physics, some evidence suggests variation is possible. Polonium radiohalos in granite suggest rapid nuclear decay. Helium retention in zircons indicates accelerated decay in the past. Perhaps during creation week or the flood, nuclear processes operated differently.

Third, the assumption of closed systems is questionable. Groundwater can add or remove isotopes, altering ratios. Metamorphism can reset isotopic clocks. The consistency of radiometric dates may reflect systematic bias rather than true age.

Young-earth creationists don't deny the existence of radioactive decay or isotope ratios. They question the interpretation of these data as indicating millions of years. Given different starting assumptions—recent creation, possible accelerated decay, and catastrophic conditions—the same data fit a young-earth timeline.

Distant Starlight

If the universe is only thousands of years old, how do we see stars millions of light-years away? This astronomical challenge requires creative solutions within the young-earth framework. Several proposals exist, each with strengths and weaknesses.

Some suggest that God created light in transit, making distant stars visible immediately. Critics note this makes God deceptive—we see stellar events that never actually happened. Others propose that the speed of light was much faster in the past, allowing distant light to reach earth quickly. This requires revising fundamental physics.

More sophisticated models invoke relativistic effects. Dr. Russell Humphreys proposes that gravitational time dilation during creation week allowed billions of years to pass in distant space while only days passed on earth. This preserves conventional physics while maintaining a young earth.

Dr. Jason Lisle's "Anisotropic Synchrony Convention" argues that the one-way speed of light is conventionally defined, not measured. By choosing a different convention, distant starlight arrives instantly. This solution is mathematically valid but conceptually challenging.

Biogeography and Distribution

Why do marsupials concentrate in Australia? Why are endemic species found on islands? Biogeography seems to support evolution and old ages. Young-earth creationists must explain how animals distributed globally from Ararat in just thousands of years.

The model proposes rapid post-flood diversification and migration. From ark kinds, numerous species emerged through built-in genetic variability. Natural selection (accepted by creationists as a conserving force) shaped populations for local environments. Migration occurred via land bridges exposed during the Ice Age when sea levels were lower.

Rafting on vegetation mats explains some distributions. Major storms can transport animals vast distances on floating debris. Laboratory experiments show many animals can survive such journeys. The post-flood world, with unstable weather and floating debris from the flood, provided ideal conditions for oceanic dispersal.

Catastrophic plate tectonics, proposed by creation geologists, suggests rapid continental movement during the flood. This could have separated populations and created biogeographic patterns. While conventional geology spreads these movements over millions of years, catastrophic conditions could accelerate the process.

Living with Tension

Young-earth creationists honestly acknowledge that their model faces scientific challenges. They don't claim to have answered every question or resolved every difficulty. Instead, they argue that starting with biblical presuppositions and working to understand data within that framework is more faithful than abandoning clear scriptural teaching for shifting scientific theories.

This approach requires living with tension. When scientific consensus contradicts their interpretation of Scripture, they hold firm to biblical authority while continuing research. They believe that further investigation will vindicate their position, just as archaeology has confirmed many biblical historical claims once doubted by scholars.

Critics charge this approach with pseudo-science—starting with conclusions and forcing evidence to fit. Young-earth creationists respond that everyone has presuppositions. Secular scientists assume naturalism and interpret evidence accordingly. The question isn't whether we have biases but whether we acknowledge them and test them against reality.

The Pastoral Dimension

Beyond scientific arguments, young-earth creationism addresses pastoral concerns. In a culture where evolution is taught as fact, many young Christians experience faith crises. If science disproves Genesis, can any of the Bible be trusted? Young-earth ministries provide resources for maintaining faith without compromising biblical authority.

The approach also maintains theological coherence. If death preceded sin, did Christ's death really defeat humanity's enemy? If humans evolved from animals, when did the image of God emerge? If Genesis is mythological, where does history begin? Young-earth creationism offers clear answers rooted in straightforward biblical interpretation.

For many believers, this clarity provides comfort. In a world of shifting values and uncertain foundations, the young-earth model offers solid ground. Genesis means what it says. God created recently. Humans are special. Sin explains suffering. Christ's redemption encompasses all creation. This comprehensive worldview appeals to those seeking certainty.

Influence and Impact

The young-earth movement has profoundly influenced evangelical Christianity. Organizations like Answers in Genesis, the Institute for Creation Research, and Creation Ministries International reach millions through publications, conferences, and attractions like the Creation Museum and Ark Encounter.

Their impact extends beyond convinced adherents. By forcing examination of assumptions and challenging evolutionary dogma, young-earth creationists have kept important questions alive. Their persistent critique of radiometric dating has prompted refinements in methodology. Their challenges to transitional fossils have spurred more careful analysis of evolutionary claims.

Critics worry about negative impacts, particularly in science education. Teaching creationism alongside evolution in schools remains contentious. Some fear that young-earth dogmatism drives scientifically-minded youth from faith. Others argue that academic freedom requires presenting alternatives to evolutionary theory.

Looking Forward

The young-earth model represents one sincere attempt to reconcile dinosaur evidence with biblical faith. Its adherents include trained scientists who sacrifice career advancement for their convictions. While facing significant scientific challenges, the model continues developing sophistication in addressing objections.

Whether dinosaurs romped with Adam in Eden or died millions of years before humans emerged, Christians must grapple with how to understand God's revelation in both Scripture and nature. The young-earth model prioritizes Scripture, reading nature through that lens. Other approaches, which we'll explore next, attempt different harmonizations.

In our next chapter, we'll examine old-earth creationism—models that accept geological time while maintaining divine creation. These approaches offer different solutions to the dinosaur dilemma, each with distinct strengths and challenges. The conversation continues as believers seek truth wherever it leads.

For Reflection

1. What aspects of the young-earth model do you find most compelling? Most challenging?
2. How do you weigh scientific consensus against biblical interpretation when they seem to conflict?
3. What role should pastoral concerns play in how we approach origins questions?

Did You Know?

The largest dinosaur eggs ever discovered are only about 18 inches long and football-shaped. Even gigantic sauropods that reached 100 feet in length started life smaller than a house cat. This biological constraint–eggs can only be so large before shells become too thick for babies to break through–supports the young-earth argument that juvenile dinosaurs could fit on Noah's ark.

Consider This

Medieval maps often included the phrase "Here be dragons" for unexplored regions. Rather than pure fantasy, young-earth creationists suggest these warnings reflected genuine encounters with surviving dinosaurs. The ubiquity of dragon legends across cultures–from St. George to Chinese dragons to Native American thunderbirds–might preserve memories of human-dinosaur interaction.

For Further Study

* Ham, Ken. The New Answers Book series
* Morris, John. The Young Earth
* Sarfati, Jonathan. Refuting Evolution
* Wood, Todd. The Quest: Exploring Creation's Hardest Problems
* Creation.com and AnswersInGenesis.org for extensive articles

Chapter 5:

"Dinosaurs Before Adam": Old-Earth Models

"For a thousand years in your sight are but as yesterday when it is past, or as a watch in the night." — Psalm 90:4

The Lawyer Who Changed Everything

Charles Lyell wasn't looking to challenge Christianity when he published Principles of Geology in the 1830s. A devout Anglican and trained lawyer, he simply wanted to establish geology as a proper science based on observable processes. His principle of uniformitarianism—"the present is the key to the past"—suggested that current geological processes, operating over vast time spans, could explain Earth's features.

For Christians who took Genesis seriously, Lyell's work posed a dilemma. The rock layers he described, the fossils they contained, and the processes that formed them seemed to require far more than 6,000 years. Yet many believers found his evidence compelling. Could there be a way to accept geological ages while maintaining biblical authority?

Long before Darwin made evolution a household word, Christian geologists and theologians were already grappling with deep time. They developed various models to harmonize Scripture with geology—approaches that accepted an ancient Earth while rejecting naturalistic evolution. These old-earth creationist models offer different solutions to the dinosaur dilemma than their young-earth counterparts, and millions of Christians today find them compelling.

The Gap Theory: A Catastrophe Before Genesis

Ruin and Reconstruction

Thomas Chalmers, Scottish Presbyterian minister and scientist, first popularized what became known as the Gap Theory in 1814. This model proposes a temporal gap between Genesis 1:1 and 1:2, during which a previous creation was destroyed. This pre-Adamic world, lasting perhaps millions of years, contained the dinosaurs and other extinct creatures found in the fossil record.

The theory hinges on translating Genesis 1:2 as "Now the earth became formless and void" rather than "was formless and void." The Hebrew word "hayah" can mean either "was" or "became," and Gap theorists argue context favors the latter. Something happened to make the earth "tohu wa bohu"—chaotic and empty.

What caused this destruction? Many Gap theorists point to Satan's fall. Isaiah 14 and Ezekiel 28, interpreted as describing Lucifer's rebellion, might chronicle this cosmic catastrophe. When Satan fell, the theory proposes, God judged the earth he had corrupted, destroying all life. The fossil record preserves this doomed world—a planet ruled by violence and death under Satan's influence.

Genesis 1:3 onward then describes not the original creation but a reconstruction. God restored the ruined planet and created new life, culminating in humanity. This explains why Genesis 1:1 uses "bara" (create from nothing) while later verses often use "asah" (make or form). The basic materials already existed from the original creation.

Solving the Dinosaur Dilemma

For Victorian Christians troubled by geological discoveries, the Gap Theory offered an elegant solution. Dinosaurs lived and died in the pre-Adamic world, explaining their absence from the biblical narrative. The rock layers and fossils formed during the undefined gap period, not during Noah's flood. This allowed acceptance of geological ages without compromising the literal six days of (re)creation.

The theory also addressed theological concerns about pre-Fall death. The violence evident in dinosaur fossils occurred in Satan's corrupted world, not God's "very good" creation. When God recreated in Genesis 1, He made a deathless paradise for humanity. Animal death before Adam thus reflected Satanic corruption, not divine intent.

C.I. Scofield popularized this view through his influential Reference Bible (1909). For decades, the Gap Theory dominated conservative Protestant thinking about origins. Dinosaurs became monuments to God's judgment on evil—proof that rebellion leads to destruction. Their fossils warned against Satan's deceptions while testifying to God's justice.

Problems and Decline

Despite its initial appeal, the Gap Theory faces significant challenges. Hebrew scholars note that the grammar of Genesis 1:2 favors "was" over "became." The phrase "formless and void" appears elsewhere in Scripture describing not judgment but pre-creation conditions awaiting God's organizing work.

The theory also creates theological difficulties. If Satan's fall precipitated global destruction, why didn't Adam's fall produce similar catastrophe? How could Satan corrupt a "very good" creation before humans sinned? The theory seems to give Satan too much power and makes evil chronologically prior to human sin.

Scientifically, the Gap Theory satisfied 19th-century geology but struggles with modern understanding. The fossil record shows not one destruction but multiple extinction events. Life's history displays progression and development, not just catastrophic judgment. The theory accepts geological time but rejects the geological narrative, an unstable compromise.

By the mid-20th century, most conservative scholars abandoned the Gap Theory. Young-earth creationism's resurgence offered a more consistent biblical literalism, while other old-earth models provided better geological correlation. Yet the theory persists in some circles, and its core insight—that Genesis might not record the absolute beginning—influences other old-earth approaches.

Day-Age Theory: When Days Aren't Days

Divine Time Versus Human Time

The Day-Age Theory takes a different approach to reconciling Scripture with science. Rather than inserting a gap, it reinterprets the days themselves. The Hebrew "yom" (day), while usually meaning a 24-hour period, can indicate longer time spans. Genesis 2:4 uses "yom" for the entire creation week. Psalm 90:4 suggests God's time perspective differs radically from ours.

If each creation "day" represents a vast age, the biblical sequence might correlate with geological history. Day One's "Let there be light" could describe the early universe becoming transparent to radiation. Day Two's separation of waters might represent atmospheric development. The progression from sea creatures to land animals to humans roughly matches the fossil sequence.

Hugh Ross, astrophysicist and founder of Reasons to Believe, champions a sophisticated version of this model. He argues that Genesis 1, properly interpreted, accurately describes 13.8 billion years of cosmic and biological history. The biblical account, written for pre-scientific readers, captures the essence of creation's unfolding while leaving details for later discovery.

Dinosaurs in Day Five and Six

In the Day-Age model, dinosaurs appear during Days Five and Six, corresponding to the Mesozoic Era. Day Five's "great sea creatures" includes marine reptiles like plesiosaurs and mosasaurs. The "birds" created this day encompass flying reptiles—pterosaurs—as well as early avians evolving from dinosaurs.

Day Six brings land dinosaurs as part of the "beasts of the earth." The sequence matches paleontology: sea life first, then land colonization. The text's categories—creatures that swarm, birds that fly, beasts that walk—describe functional groups rather than modern taxonomic classifications. Dinosaurs fit naturally within this framework.

Their extinction before humanity also makes sense in this model. Each "day" involves both creation and removal of creatures as God prepares Earth for its ultimate purpose—hosting beings bearing His image. Dinosaurs served their role in Earth's development but gave way to mammals when conditions changed. Their 160-million-year reign was just preparation for humanity's arrival.

This view handles the death-before-sin problem by distinguishing animal from human death. Romans 5:12 speaks specifically of human death through sin. Animals, lacking eternal souls and moral agency, experienced physical death as natural limitation. The "very good" creation included cycles of life and death that maintained ecological balance. Only humans were created for immortality, lost through sin.

Challenges to Day-Age Interpretation

Critics from both young-earth and secular perspectives challenge the Day-Age model. Young-earthers argue it compromises biblical authority by reinterpreting clear language. The phrase "evening and morning" seems to indicate normal days. The Fourth Commandment loses coherence if God's work week lasted billions of years.

The order of events also poses problems. Genesis has fruit trees before sea creatures, birds before land animals, and light before the sun. While Day-Age proponents offer explanations—describing Earth-surface perspective rather than cosmic chronology—critics find these harmonizations forced.

Scientifically, the model requires significant concordism—

reading modern science into ancient texts. Does Genesis really describe the Big Bang, stellar evolution, and plate tectonics? Or are interpreters imposing contemporary knowledge on writings with different purposes? The danger is making Genesis say things Moses' audience couldn't possibly understand.

Progressive Creation: God's Unfolding Plan

Creation in Stages

Progressive Creation offers another old-earth perspective, proposing that God created life in discrete stages over geological time. Unlike theistic evolution, this view maintains that God directly intervened at key points—creating major new forms rather than merely guiding natural processes.

Bernard Ramm's The Christian View of Science and Scripture (1954) articulated this position for evangelicals. He argued that the "kinds" of Genesis correspond roughly to major taxonomic groups. God created these archetypal forms at appropriate times throughout Earth history. Within kinds, variation and adaptation occur, but new kinds require divine action.

This model sees the fossil record's apparent progression not as evolution but as progressive creation. The Cambrian Explosion, where major animal phyla appear suddenly, exemplifies God's creative acts. Each geological era witnesses new creations as God prepares Earth for its ultimate inhabitants. Dinosaurs represent one stage in this divine plan.

The days of Genesis, in this view, may be either long ages or 24-hour days marking God's creative acts. Some progressive creationists see the days as visions given to Moses—literal days of revelation about ages of creation. Others view them as a logical framework explaining creation's order rather than chronology.

Divine Purpose in Deep Time

Why would God take billions of years to create? Progressive creationists see deep time as purposeful, not problematic. Each stage prepared Earth for the next. Microscopic life oxygenated the atmosphere. Plants created soil. Dinosaurs processed vegetation and distributed nutrients. Each creature played its role in making Earth habitable for humanity.

This perspective finds theological richness in Earth's long history. God's patience and providence shine through eons of preparation. The same care that crafted dinosaurs' complex ecosystems prepared every detail for human flourishing. Time itself becomes a canvas for divine artistry.

Dinosaurs gain special significance in this model. Their size and power demonstrate God's creativity unconstrained by present limitations. Their extinction shows divine sovereignty over life and death. Their fossils provide resources—coal and oil—for human civilization. Even extinct creatures serve God's purposes.

Progressive creation also addresses biogeography better than young-earth models. The distribution of species reflects long histories in specific locations. Australia's marsupials, Madagascar's lemurs, and the Galápagos' finches result from extended divine activity in particular places. God crafted each ecosystem over time for His purposes.

Distinguishing from Evolution

Progressive creationists strongly distinguish their view from theistic evolution. They reject universal common descent, insisting that major groups require special creation. While accepting microevolution within kinds, they deny macroevolution between kinds. Genetic similarities reflect common design, not common ancestry.

The model maintains clear discontinuities in life's history. The origin of life required divine action—naturalistic chemical evolution cannot explain life's beginning. Major transitions—prokaryotes to eukaryotes, single-celled to multicellular, invertebrates to vertebrates—needed God's direct intervention. Dinosaurs to birds might involve guided development within a kind, but reptiles to mammals required new creation.

Human creation especially demands divine action. While progressive creationists may accept anatomical similarities between humans and primates, they insist the image of God came through special creation. Whether God created humans de novo or transformed existing hominids, the result was qualitatively different—beings with souls, moral agency, and spiritual capacity.

Framework Interpretation: Literary Rather Than Literal

Reading Genesis as Ancient Literature

Some old-earth creationists approach Genesis 1 primarily as theological literature rather than scientific description. The Framework Interpretation, developed by Reformed scholars like Meredith Kline, sees the creation week as a literary structure conveying theological truth rather than chronological sequence.

The days form two parallel triads. Days 1-3 create realms (light/darkness, sky/sea, land). Days 4-6 fill these realms (luminaries, birds/fish, land animals/humans). This literary symmetry suggests artistic arrangement rather than historical narrative. The author crafted a memorable framework to teach creation theology.

This view liberates Genesis from scientific correlation. The text teaches that God created everything, not how or when. Questions about dinosaurs become scientifically rather than biblically determined. Genesis addresses who and why; science investigates what and when. Neither threatens the other when properly understood.

Ancient Near Eastern context supports this reading. Creation accounts from surrounding cultures used similar literary devices. Genesis 1 critiques these pagan myths while employing familiar forms. The ordered structure counters chaotic pagan cosmogonies. The repeated "good" proclamations challenge dualistic worldviews. Read this way, Genesis is theology in literary dress.

Implications for Understanding Dinosaurs

If Genesis 1 provides theological framework rather than scientific detail, dinosaurs fit wherever science places them. Their existence glorifies God without requiring biblical mention. The text's silence about countless creatures—from bacteria to beetles—doesn't diminish their witness to divine creativity.

This approach allows full acceptance of paleontological findings. Dinosaurs lived when geology indicates, evolved as evidence suggests (within limits), and went extinct through natural processes. Yet each discovery reveals God's handiwork. Science becomes a means of exploring creation's wonders without threatening biblical authority.

The Framework view especially appeals to Christians in scientific fields. They can pursue research without constantly reconciling findings with concordist interpretations. A paleontologist can study dinosaur evolution while worshiping their Creator. A geologist can accept radiometric dates while affirming divine sovereignty.

Comparing Old-Earth Models

Each old-earth model offers distinct advantages and faces unique challenges. The Gap Theory preserves literal creation days but requires questionable Hebrew grammar and creates theological difficulties. Day-Age maintains creation sequence but struggles with textual details and concordist temptations. Progressive Creation affirms divine involvement but must explain discontinuities in the fossil record. Framework Interpretation respects literary genre but risks divorcing theology from history.

What unites these approaches is acceptance of mainstream geological timing while maintaining divine creation. They agree dinosaurs preceded humans by millions of years. They concur that Genesis teaches theology, not technical science. They affirm that nature and Scripture, properly interpreted, cannot conflict.

These models also share common critiques of young-earth creationism. They find attempts to compress Earth history into thousands of years scientifically untenable. They see flood geology as inadequate to explain geological features. They believe young-earth teaching, however well-intentioned, damages Christian credibility and creates unnecessary barriers to faith.

The Appeal of Old-Earth Creationism

Why do many Christians find old-earth models compelling? First, they take mainstream science seriously. Rather than dismissing centuries of geological research, these models seek integration. Christians in scientific fields especially appreciate approaches that don't require rejecting their disciplines' foundations.

Second, old-earth models often provide richer theology. Deep time reveals God's patience, wisdom, and sovereignty in ways young-earth timelines cannot. The long preparation for humanity demonstrates divine purpose and planning. Each extinct creature, including dinosaurs, played its role in God's grand design.

Third, these approaches may remove obstacles to faith. In cultures where evolution is assumed, young-earth creationism can seem intellectually impossible. Old-earth models offer stepping stones for those journeying toward faith. They demonstrate that serious Christians can accept scientific findings while maintaining biblical faith.

Finally, old-earth creationism emphasizes humility. Given nature's complexity and Scripture's depth, dogmatism seems inappropriate. These models acknowledge mysteries and uncertainties while affirming core truths. They major on majors—God as Creator—while allowing freedom on details.

Continuing Challenges

Despite their appeal, old-earth models face ongoing challenges. Theologically, they must explain pre-Fall animal death satisfactorily. While distinguishing human from animal death helps, the presence of suffering and extinction before sin remains troubling for some.

Biblically, these models risk weakening scriptural authority. If Genesis 1 doesn't mean what it plainly says, how do we determine what texts to take literally? The hermeneutical principles that allow reinterpreting creation days might be applied to other passages, potentially undermining essential doctrines.

Scientifically, old-earth creationists must define their acceptance of evolutionary theory. Where exactly does divine intervention occur? How do we detect special creation scientifically? The models can seem like "God of the gaps" reasoning, invoking divine action wherever natural processes seem insufficient.

Pastorally, these approaches may confuse laypeople. Young-earth creationism offers clear, simple answers. Old-earth models require nuanced thinking about Hebrew grammar, ancient literary genres, and scientific methodology. Not everyone has the background to evaluate these complex arguments.

Living with Tension

Old-earth creationists often describe living with creative tension. They hold firm convictions about God as Creator while remaining open about creation's details. They worship the God revealed in Scripture while marveling at nature's revelations. They engage both ancient texts and modern science with reverence and rigor.

This tension can be fruitful. It drives deeper Bible study and better science. It fosters humility and wonder. It creates space for dialogue between Christians with different views. It acknowledges that finite humans investigating infinite God's works will inevitably encounter mystery.

Norman Geisler, a prominent old-earth creationist, argued that the age of Earth is not a test of orthodoxy. Christians can disagree about when God created while agreeing that He created. The gospel doesn't depend on settling the dinosaur debate. Christ's resurrection matters infinitely more than Tyrannosaurus's extinction date.

Bridge to Design

Old-earth creationism opens important questions about detecting divine action in nature. If God created progressively over millions of years, can we identify His fingerprints? Do dinosaurs and other creatures display evidence of design that transcends natural processes?

Our next chapter explores the Intelligent Design movement, which seeks to answer these questions. While not exclusively old-earth, ID proponents often accept geological timescales while arguing that life's complexity requires intelligent causation. Their approach offers yet another perspective on understanding dinosaurs within a theistic framework.

Whether dinosaurs lived before Adam in a previous creation, developed through divinely guided ages, or appeared through progressive divine acts, old-earth creationists affirm their witness to God's power and wisdom. These ancient creatures still speak, declaring their Maker's glory across whatever timespan He employed.

For Reflection

1. Which old-earth model do you find most persuasive? What are its strongest and weakest points?
2. How do you understand the relationship between animal death before the Fall and human death after it?
3. What criteria should guide Christians in interpreting Genesis–literary, theological, scientific, or traditional?

Did You Know?

The idea of an old Earth predates Darwin by decades. Christian geologists like William Buckland and Adam Sedgwick accepted deep time based on geological evidence while completely rejecting evolution. They saw vast ages as magnifying God's glory, not diminishing it. The conflict between geology and Genesis troubled believers long before evolution entered the picture.

Consider This

Dimetrodon, often mistaken for a dinosaur, actually lived before dinosaurs appeared and was more closely related to mammals. This sail-backed predator from the Permian period (about 295-272 million years ago) reminds us that Earth's history included multiple chapters of spectacular creatures. Each era–including the age of dinosaurs–prepared the way for the next in God's unfolding plan.

For Further Study

* Ross, Hugh. A Matter of Days and Navigating Genesis
* Ramm, Bernard. The Christian View of Science and Scripture
* Geisler, Norman. Systematic Theology, Volume 2
* Collins, C. John. Genesis 1-4: A Linguistic, Literary, and Theological Commentary
* Young, Davis. The Biblical Flood: A Case Study of the Church's Response to Extrabiblical Evidence

Chapter 6:

"Intelligent Design": Finding the Designer's Fingerprints

"The heavens declare the glory of God, and the sky above proclaims his handiwork." — Psalm 19:1

The Professor Who Saw Too Much

Michael Behe was a comfortable evolutionary biologist at Lehigh University when a colleague loaned him Michael Denton's Evolution: A Theory in Crisis. Behe expected to find easily refutable creationist arguments. Instead, he encountered detailed biochemical challenges to Darwinian evolution that he couldn't dismiss. The more he investigated, the more convinced he became that life's molecular machinery couldn't arise through undirected processes.

In 1996, Behe published Darwin's Black Box, introducing the concept of "irreducible complexity" to public discourse. A mousetrap, he explained, requires all its parts to function—remove any component and it fails completely. Similarly, molecular machines like the bacterial flagellum need all their parts present and properly arranged. How could gradual evolution produce systems that only work when complete?

This launched what became known as the Intelligent Design (ID) movement, though its intellectual roots stretch back through William Paley's watchmaker argument to Aristotle's prime mover. ID proponents claim to detect empirical evidence of design in nature while remaining officially agnostic about the designer's identity. For dinosaurs, this approach offers fascinating perspectives on their remarkable engineering.

The Design Revolution

Beyond Creationism

Intelligent Design differs significantly from traditional creationism. While most ID proponents are theists, the movement includes agnostics and even some unconventional atheists. They unite around a central claim: certain features of life and the universe exhibit patterns best explained by intelligent causation rather than undirected processes.

ID advocates emphasize their approach is scientific, not religious. They don't start with Scripture or theological commitments but with empirical observation. Complex specified information, irreducible complexity, and fine-tuning point to intelligence. Whether that intelligence is God, aliens, or something else lies beyond ID's scientific scope—though most supporters privately identify the designer with the God of theism.

This distinction matters legally and educationally. ID proponents argue their theory belongs in science classrooms because it makes scientific claims based on evidence, not religious doctrine. Critics counter that ID is creationism in disguise, but supporters insist the approach differs fundamentally. Creationism starts with revealed truth and interprets evidence accordingly; ID starts with evidence and infers design.

The movement gained institutional support through Seattle's Discovery Institute, founded in 1990. Their Center for Science and Culture promotes ID research and education. Key figures include Behe, mathematician William Dembski, philosopher Stephen Meyer, and biologist Jonathan Wells. Together they've challenged evolutionary orthodoxy while developing sophisticated arguments for design.

Dinosaurs as Design Marvels

When ID theorists examine dinosaurs, they see engineering marvels that defy gradual evolutionary explanation. Consider

Pteranodon, the giant pterosaur with a twenty-foot wingspan. Everything about this creature seems optimized for flight—hollow bones for lightness, elongated fourth finger supporting the wing membrane, specialized shoulder joints for powered flight, and a brain adapted for three-dimensional navigation.

The challenge for evolution is explaining how land reptiles developed flight. The transitional forms would seemingly be disadvantaged—proto-wings too small to fly would hinder running and climbing. Yet Pteranodon shows no evidence of compromise. Its anatomy appears perfectly designed for its aerial lifestyle. ID proponents see this as evidence of intelligent planning rather than blind selection.

Or consider the ankylosaur's defensive systems. Euoplocephalus possessed coordinated armor plating, a bone-crushing tail club, and even protected eyes. Each element alone provides some benefit, but together they form an integrated defense system. The bone structure, muscle attachments, and neural control needed for an effective tail club weapon suggest forethought and design.

Even more remarkable are dinosaur behaviors preserved in the fossil record. Maiasaura nesting colonies show evidence of parental care rivaling modern birds. Adults apparently brought food to helpless young, protected nests from predators, and may have taught survival skills. Such complex behaviors require sophisticated neural programming. How did unguided mutations produce the integrated anatomical and behavioral systems necessary for successful parenting?

The Problem of Convergent Evolution

Convergent evolution—similar features arising independently in unrelated lineages—particularly troubles evolutionary theory while supporting design. Consider the uncanny similarities between Deinonychus (a dinosaur) and Thylacoleo (an extinct marsupial predator). Both developed sickle-shaped killing claws, powerful jaw muscles, and similar hunting strategies despite sharing no recent common ancestor.

From an ID perspective, convergence reflects design principles applied to similar problems. Just as human engineers independently develop similar solutions—multiple cultures invented the wheel—an intelligent designer might use optimal designs repeatedly. The prevalence of convergence in nature suggests underlying design patterns rather than random exploration of possibilities.

Dinosaur eyes provide another example. Large theropods like Tyrannosaurus developed forward-facing eyes with binocular vision for depth perception—crucial for predators. The same feature appears in completely unrelated predatory mammals. Random mutations should explore different solutions, but we see the same sophisticated visual system arising repeatedly. This suggests design constraints or templates rather than undirected variation.

Complexity That Darwin Never Imagined

Molecular Machines in Ancient Creatures

When Darwin proposed natural selection, the cell was a mysterious blob of protoplasm. Today's biochemistry reveals molecular machines of staggering complexity. While we can't directly study dinosaur biochemistry, their living relatives—birds and crocodiles—provide insights into the molecular marvels dinosaurs likely possessed.

The avian respiratory system exemplifies irreducible complexity. Birds have flow-through lungs unlike the bellows-style lungs of mammals. Air flows continuously in one direction through a system of air sacs and lungs, extracting oxygen far more efficiently than mammalian respiration. This system requires precisely coordinated anatomy—specialized bones, air sacs, and lung structure—that only functions when complete.

Evidence suggests some dinosaurs shared this respiratory system. Theropod dinosaurs show pneumatic (air-filled) bones characteristic of avian-style respiration. Aerosteon, discovered in Argentina, preserves clear evidence of air sacs throughout its body. This efficient breathing system may have enabled the high activity levels seen in many dinosaurs.

But how could such a system evolve gradually? The intermediate stages between bellows lungs and flow-through lungs would seemingly be less efficient than either complete system. The extensive anatomical reorganization required—rerouting airways, developing air sacs, modifying bones—must happen in coordinated fashion. ID proponents argue this points to deliberate design rather than step-by-step evolution.

Information Systems in Stone

Perhaps the most powerful ID argument concerns biological information. DNA contains coded instructions for building organisms, genuine digital information comparable to computer code. For dinosaurs, we're beginning to glimpse this information system through exceptional preservation.

Mary Schweitzer's discovery of soft tissue in dinosaur fossils opened unprecedented windows into molecular paleontology. Her team has recovered protein fragments from multiple dinosaur species. These proteins show complex sequences that determine their three-dimensional structure and function. Each protein represents thousands of bits of specified information.

The origin of biological information poses evolution's greatest challenge. Random mutations might modify existing information but struggle to create new functional sequences. It's like expecting random letter changes to transform one novel into another—possible in theory but vanishingly improbable. The proteins found in dinosaur tissues represent information that must originate somewhere.

ID theorists argue that information always traces back to intelligent sources. We recognize design in human artifacts—books, computer programs, buildings—because they contain complex specified information. Similarly, the genetic information that built Triceratops's three-horned skull or Stegosaurus's plate arrangement points to an intelligent source. Natural processes might shuffle existing information but cannot explain its origin.

The Cambrian Explosion Connection

Life's Big Bang

While dinosaurs appeared much later, the Cambrian Explosion provides crucial context for ID arguments. About 540 million years ago, most major animal body plans appeared suddenly in the fossil record. This "biological big bang" produced the fundamental architectures that all subsequent animals, including dinosaurs, would modify.

Stephen Meyer's Darwin's Doubt argues the Cambrian Explosion exemplifies intelligent design. The rapid appearance of complex animals with integrated organ systems, sophisticated eyes, and ecological relationships defies gradual evolutionary explanation. Where are the Precambrian ancestors? Why did evolution suddenly accelerate? How did so much biological information originate so quickly?

The pattern resembles human technological development more than natural evolution. New technologies often appear suddenly when inventors combine existing knowledge in novel ways. The Cambrian Explosion might represent something similar—an intelligent agent introducing new biological designs. These foundational body plans provided templates that later creatures, including dinosaurs, would elaborate.

Dinosaurs themselves emerged somewhat suddenly in the Triassic period. The earliest dinosaurs like Eoraptor already possessed the distinctive dinosaurian body plan—upright stance, specialized hip structure, and particular ankle anatomy. The transition from primitive archosaurs to true dinosaurs happened rapidly with few clear intermediates. This pattern of sudden appearance followed by variation within types matches ID predictions better than gradual evolution.

Design Patterns in Deep Time

ID proponents see patterns throughout life's history that suggest intentional planning. Mass extinctions followed by explosive diversification look like designed resets rather than random catastrophes. The progression from simple to complex follows engineering logic—establishing basic systems before adding sophistication.

Consider how dinosaurs fit this pattern. They appeared after the Permian extinction cleared ecological space. Their diversification filled specific niches with appropriate designs—long-necked sauropods for high browsing, armored ankylosaurs for defense, swift ornithomimids for pursuit predation. The efficiency with which dinosaurs radiated into available ecological roles suggests pre-existing design plans activated when conditions permitted.

The replacement of dinosaurs by mammals follows similar patterns. Mesozoic mammals remained small and ecologically marginal while dinosaurs dominated. After dinosaur extinction, mammals exploded into vacant niches with remarkable speed. This coordinated replacement—one dominant group succeeding another—hints at orchestrated succession rather than random replacement.

Flight: The Ultimate Design Challenge

Conquering the Skies

Powered flight represents one of nature's greatest engineering achievements, requiring integrated solutions to multiple challenges. Dinosaurs and their relatives achieved flight at least three times—pterosaurs, birds, and possibly some small theropods like Microraptor. Each solution shows sophisticated design that challenges gradual evolution.

Quetzalcoatlus, the largest pterosaur, had a forty-foot wingspan yet weighed perhaps 500 pounds. Its bones were hollow but reinforced at stress points. Wing membranes attached precisely to allow controlled flight. The brain possessed specialized regions for processing three-dimensional movement. Everything about this creature was optimized for flight despite its massive size.

How does evolution produce flight? The traditional scenario envisions small dinosaurs leaping from trees, with proto-wings providing slight gliding advantages. But functional analysis reveals problems. Partial wings hinder climbing and running. True powered flight requires simultaneous changes in skeleton, muscles, nervous system, and metabolism. The transitional forms should be inferior to both terrestrial and aerial specialists.

Archaeopteryx, the famous "missing link," actually highlights design features rather than solving them. This creature possessed fully formed flight feathers with asymmetric vanes and intricate barbule systems identical to modern birds. These feathers represent irreducibly complex structures—remove any component and they fail. Yet Archaeopteryx appears suddenly with perfect feathers, not primitive proto-feathers.

The recent discovery of four-winged dinosaurs like Microraptor complicates evolutionary scenarios while supporting design. Why would evolution experiment with four-winged flight when two wings work fine? From a design perspective, different solutions to flight reflect creative exploration of possibilities. An intelligent designer might implement multiple flight strategies to fill different niches.

Feathers: Nature's High Technology

Feathers deserve special attention as marvels of engineering. A single flight feather contains millions of precisely arranged components. The central rachis provides strength while remaining lightweight. Barbs branch from the rachis at optimal angles.

Barbules extend from barbs with tiny hooks that zip together, creating a continuous surface that can be repaired by preening.

This hierarchical structure from macro to micro scale resembles human engineering more than biological accidents. Modern engineers study feathers for insights into lightweight, self-repairing materials. The fact that such sophisticated structures appear fully formed in the earliest feathered dinosaurs suggests design rather than gradual development.

Recent discoveries of feathered dinosaurs in China reveal additional complexity. Different dinosaurs possessed different feather types—simple filaments, downy plumes, symmetric decorative feathers, and asymmetric flight feathers. This diversity appearing simultaneously challenges linear evolution while matching expectations of creative design exploring multiple solutions.

Dinosaur Behavior and Instinct

Programmed for Success

Fossilized behavior provides unique evidence for design. Oviraptor fossils found brooding nests died protecting their eggs, arms spread in protective posture. This behavior required integrated systems—recognition of eggs as valuable, instinctive positioning, willingness to sacrifice for offspring. How did undirected mutations produce such sophisticated behavioral programming?

Pack hunting in dinosaurs like Deinonychus suggests complex social intelligence. Coordinated hunting requires communication, role differentiation, and tactical planning. The fossil site that inspired Jurassic Park's raptors shows multiple individuals who died attacking a much larger Tenontosaurus. This implies sophisticated neural programming for cooperative behavior.

Migration patterns preserved in dinosaur bone chemistry indicate some species traveled hundreds of miles seasonally. This behavior demands navigation abilities, environmental awareness, and physiological adaptations for long-distance travel. The integrated systems required—sensory apparatus, neural processing, metabolic efficiency—suggest deliberate design rather than accidental accumulation of features.

Responding to Critics

The God of the Gaps?

Critics charge ID with "god of the gaps" reasoning—invoking supernatural explanation wherever natural causes seem insufficient. ID proponents respond that they're not arguing from ignorance but from positive evidence of design. We recognize design through positive indicators—complex specified information, irreducible complexity, fine-tuning—not merely from evolutionary failures.

Furthermore, ID makes testable predictions. If life is designed, we expect to find irreducibly complex systems, optimal solutions to engineering problems, and convergent design patterns. These predictions are confirmed repeatedly in dinosaur studies. The efficiency of dinosaur designs, from T. rex's bone-crushing bite to Diplodocus's weight-bearing architecture, matches design expectations.

Methodological Naturalism

Science, critics argue, must assume methodological naturalism—explaining phenomena through natural causes alone. ID violates this principle by introducing intelligent causation. ID advocates respond that this artificially limits science. If intelligent causes exist and produce empirically detectable effects, why exclude them a priori?

Consider how we study dinosaur behavior. When paleontologists find Maiasaura nests arranged in colonies with evidence of parental care, they infer intelligent behavior. When we discover Triceratops herds with juveniles protected in the center, we recognize purposeful organization. Intelligence is already part of scientific explanation—ID simply extends this to life's origin and development.

The Identity Question

Who is the designer? ID officially remains agnostic, focusing on detecting design rather than identifying designers. Critics see this as disingenuous—clearly most ID proponents believe in God. But the distinction has merit. SETI researchers look for intelligent signals without knowing sender identity. Archaeologists identify artifacts as designed before determining their makers.

For dinosaurs, the design patterns point to an intelligence operating across vast time scales with comprehensive biological knowledge. The designer(s) must preexist Earth life, possess staggering power and intellect, and work through extended time periods. While ID science can't specify further, these characteristics certainly compatible with the God of biblical theism.

Dinosaurs in the Design Paradigm

Within ID framework, dinosaurs represent spectacular examples of biological engineering. Their size pushed physical limits, requiring innovative solutions. Sauropod necks needed special valves preventing blood pooling. Theropod legs balanced massive bodies for bipedal running. Hadrosaur dental batteries processed tough vegetation efficiently. Each innovation shows hallmarks of intelligent problem-solving.

The diversity of dinosaur designs reflects creative exploration within constraints. Just as human engineers produce variations on themes—different car models sharing basic architecture—dinosaurs show theme and variation. All theropods share basic body plans but express amazing diversity. This matches expectations of intelligent design working within established parameters.

Extinction doesn't negate design. Human technologies become obsolete; buildings are demolished. Similarly, dinosaur designs served their purpose in their time. Their extinction cleared ecological space for mammalian radiation, perhaps part of larger design strategy. The fossil record might chronicle not evolutionary meandering but purposeful biological succession.

Future Directions

ID research on dinosaurs continues expanding. New analytical techniques allow deeper investigation of preserved soft tissues. Computer modeling tests the engineering efficiency of dinosaur anatomy. Information theory quantifies the complexity of biological systems. Each advance tends to reveal greater sophistication rather than evolutionary simplicity.

The movement faces ongoing challenges. Legal battles over teaching ID continue. Many scientists remain hostile to design inferences. The reluctance to identify the designer frustrates both religious supporters and secular critics. Yet ID has succeeded in raising fundamental questions about life's complexity that demand answers.

For Christians interested in dinosaurs, ID offers valuable perspectives without requiring specific interpretations of Genesis. Whether God created through direct action, progressive creation, or guided evolution, the evidence of design remains. Dinosaurs declare their Maker's glory through their magnificent engineering, regardless of creation's timeline or mechanism.

Looking Ahead

Our next chapter examines theistic evolution—the view that God created through evolutionary processes. This perspective differs sharply from ID by fully accepting evolutionary theory while maintaining divine purpose. How do theistic evolutionists understand dinosaurs? Can evolution itself be God's design method? These questions take us into one of today's most active areas of science-faith dialogue.

The debate between ID and theistic evolution often divides along disciplinary lines. Mathematicians and engineers tend toward ID, impressed by life's information content and engineering. Biologists often prefer theistic evolution, seeing God's hand in evolution's creative power. Both seek to honor God while taking science seriously—they simply disagree on how design and natural processes interact.

For Reflection

1. Do you find the concept of irreducible complexity convincing when applied to dinosaur features? Why or why not?
2. How does detecting design in nature affect your view of God's relationship to creation?
3. Is the ID movement's refusal to identify the designer a strength or weakness?

Did You Know?

The bacterial flagellum, ID's poster child for irreducible complexity, spins at up to 100,000 rpm and can reverse direction in a quarter turn. This molecular motor contains about 40 protein parts that must mesh precisely. Some scientists argue similar molecular machines likely powered dinosaur cells, representing nanotechnology that human engineers are only beginning to match.

Consider This

Dracorex hogwartsia (the "Dragon King of Hogwarts") had a skull so dramatically spiky and dragon-like that scientists gave it a fanciful name. From an ID perspective, such apparently "excessive" features challenge purely functional evolution. Why develop such elaborate display structures? Perhaps they reflect a designer's artistic flair beyond mere survival requirements–beauty for its own sake.

For Further Study

* Behe, Michael. Darwin's Black Box and Darwin Devolves
* Meyer, Stephen. Darwin's Doubt and Signature in the Cell
* Dembski, William. The Design Inference
* Bechly, Günter. Various papers on fossil discontinuities
* Discovery Institute website: discovery.org/id

Chapter 7:

"God Created Through Evolution": Theistic Evolution

"When I look at your heavens, the work of your fingers, the moon and the stars, which you have set in place, what is man that you are mindful of him?" — Psalm 8:3-4

The Geneticist Who Found God in DNA

Francis Collins was an atheist when he entered medical school. The son of freethinkers, he considered faith an outdated superstition incompatible with scientific rationality. But conversations with dying patients who drew strength from their beliefs unsettled his confident materialism. How could he dismiss what gave such profound meaning and peace?

Reading C.S. Lewis's Mere Christianity began Collins's journey from atheism to faith. The moral law argument particularly resonated—why do humans universally sense right and wrong if we're merely molecular machines? By age 27, this rising scientist knelt and surrendered to Christ, beginning a lifelong integration of faith and science.

Collins went on to lead the Human Genome Project, one of history's great scientific achievements. In 2006, he published The Language of God, arguing that DNA itself reveals divine authorship. Evolution, rather than threatening faith, becomes God's elegant method of creation. This perspective—theistic evolution or evolutionary creation—offers another lens for understanding dinosaurs and their place in God's plan.

Evolution as God's Method

The Harmony of Two Books

Theistic evolutionists embrace both Scripture and nature as revelations from God. The Bible reveals who created and why; science discovers how and when. These "two books" cannot ultimately conflict because they have the same Author. Apparent contradictions stem from misinterpreting either Scripture or nature, not from fundamental opposition.

This approach differs radically from both atheistic evolution and special creationism. Unlike atheists, theistic evolutionists see purpose, direction, and divine sustenance throughout evolution. Unlike special creationists, they accept mainstream scientific accounts of life's development. God doesn't intervene periodically to create new species—He crafted natural processes that unfold His intentions across deep time.

BioLogos, founded by Collins, articulates this view for evangelical audiences. They affirm core Christian doctrines—God as Creator, human uniqueness, biblical authority—while accepting evolutionary science. Dinosaurs fit naturally within this framework as one stage in God's creative process, preparing Earth for humanity's eventual arrival.

The model addresses a crucial pastoral concern. Many young Christians lose faith when forced to choose between science and Scripture. Theistic evolution offers a third way, preserving both scientific integrity and spiritual devotion. Students can study paleontology without fearing for their souls, seeing each fossil as revealing God's patient artistry.

Dinosaurs in the Divine Drama

From a theistic evolutionary perspective, dinosaurs played essential roles in God's unfolding creation. Their 165-million-year reign wasn't pointless meandering but purposeful preparation. Through dinosaurs, God crafted ecosystems, distributed nutrients globally, and developed biological innovations that would benefit later creatures.

Consider how dinosaurs shaped Earth's biology. Sauropods, processing tons of vegetation daily, were living nutrient distributors. Their droppings fertilized vast areas, while their trampling created habitat diversity. Predatory dinosaurs maintained ecological balance, preventing any species from overwhelming others. This dynamic ecosystem prepared soil, climate, and biological foundations for mammalian diversification.

The extinction of dinosaurs further demonstrates divine providence. The asteroid impact 66 million years ago cleared ecological space for mammals, our ancestors. What seems catastrophic was actually creative, opening possibilities for new forms. God used even cosmic collisions to advance His purposes. Evolution's apparent randomness masks deeper purpose visible only in hindsight.

Convergent evolution, problematic for purely naturalistic accounts, makes perfect sense within theistic evolution. When similar environments pose similar challenges, God's evolutionary processes produce similar solutions. The independent evolution of camera eyes in vertebrates and cephalopods reflects not coincidence but divine consistency. God embedded potential solutions within creation's fabric.

The Role of "Randomness"

Critics often ask how random mutations can serve divine purposes. Doesn't evolution's dependence on chance eliminate God's sovereignty? Theistic evolutionists respond with nuanced theology about providence and probability.

First, "random" in science means unpredictable by natural laws, not purposeless. When geneticists say mutations occur randomly, they mean we can't predict which DNA base will change when. This doesn't preclude divine influence any more than casting lots (Proverbs 16:33) prevents God's sovereignty. God can work through apparently random processes.

Second, evolution isn't purely random—natural selection provides direction. Beneficial mutations survive; harmful ones disappear. This filtering process, designed by God, guides evolution toward greater complexity and adaptation. The system's elegance lies in achieving specific outcomes through general processes. God doesn't micromanage each mutation but designed laws that inevitably produce intended results.

Third, quantum mechanics suggests reality's fundamental indeterminacy. If God influences quantum events, He could guide evolution without violating natural laws. Divine action might be scientifically undetectable yet theologically certain. This isn't "god of the gaps" but recognition that science describes mechanisms, not ultimate causes.

The Cambrian Explosion Reconsidered

Creative Explosion, Not Creation Ex Nihilo

The Cambrian Explosion poses interesting questions for theistic evolution. How do evolutionary creationists explain the apparently sudden appearance of complex animals? Rather than seeing this as evidence against evolution, they view it as evolution in overdrive—God's creativity bursting forth when conditions aligned.

Recent discoveries reveal the Cambrian Explosion was less sudden than previously thought. Precambrian fossils show earlier complexity—bilateral symmetry, developmental toolkits, ecological relationships. The "explosion" represents fossilization conditions improving as much as evolution accelerating. When animals developed hard parts, they left better records.

Theistic evolutionists see God preparing for the Cambrian throughout the preceding Ediacaran period. Genetic regulatory systems evolved that could generate diverse body plans. Oxygen levels rose to support active metabolisms. Ecological niches opened as microbes declined. When these factors converged, evolution rapidly explored new possibilities—not randomly but following potentials God embedded in creation.

This perspective celebrates scientific discoveries rather than fearing them. Each Precambrian fossil that "softens" the Cambrian Explosion reveals more of God's patient preparation. The process becomes more wonderous, not less, as we understand its complexity. God's creativity works through natural processes without being limited by them.

From Cambrian Foundations to Dinosaur Innovations

The body plans established during the Cambrian provided templates for all subsequent animals, including dinosaurs. Chordates—our phylum—appeared with notochords, dorsal nerve cords, and other features that would eventually support massive sauropods and agile theropods. God's Cambrian creativity established frameworks allowing endless variation.

Dinosaurs represent spectacular variations on these themes. The basic reptilian body plan, refined through millions of years, produced forms pushing physical limits. Argentinosaurus grew to 100 tons by optimizing bone structure and metabolism. Microraptor achieved powered flight through incremental feather modifications. Each innovation built on previous foundations in ways that seem guided despite appearing natural.

This cumulative creativity reflects how God typically works—through processes, not magic. Jesus compared the Kingdom of God to slowly growing seeds, not instant gardens. Creation follows similar patterns, with patient development yielding eventual glory. Dinosaurs emerged through eons of gradual change, but this makes them no less divinely intended.

Human Uniqueness in an Evolutionary World

The Image of God Question

The most challenging aspect of theistic evolution concerns human origins. If humans evolved from earlier primates, when and how did we receive God's image? How do we reconcile evolutionary continuity with biblical teaching about human uniqueness? Dinosaurs, extinct before these questions arise, provide helpful perspective on the broader issues.

Different theistic evolutionists propose various solutions. Some see the image of God as emergent—when brains reached sufficient complexity, consciousness, morality, and spirituality became possible. Others propose God specially intervened to endow evolved hominids with souls. Still others interpret the image functionally—humans appointed as creation's representatives regardless of origins.

What's clear from the dinosaur fossil record is that physical form alone doesn't determine value or purpose. Triceratops had larger brains than early mammals, yet didn't bear God's image. Velociraptors showed social intelligence but lacked moral agency. The image of God transcends biology while working through it. However humans acquired this image, it represents something qualitatively new in creation.

The long prehistory before humans—including the age of dinosaurs—demonstrates God's patience and purpose. He spent billions of years preparing Earth for image-bearers. This wasn't inefficiency but artistry, like a master chef preparing ingredients before combining them perfectly. Dinosaurs helped create the world humans would eventually inhabit and rule.

Death Before the Fall Revisited

Theistic evolution requires accepting animal death before human sin. The fossil record, including dinosaur graveyards, documents millions of years of predation, disease, and extinction.

How does this square with creation being "very good"?

Many theistic evolutionists distinguish between human death (spiritual and physical) and animal mortality. Romans 5:12 specifically addresses human death through sin. Animals, lacking eternal souls and moral culpability, experienced physical death as natural limitation. The "very good" creation included cycles maintaining balance and driving adaptation.

Furthermore, death and renewal characterize all life. Seeds "die" to produce plants. Cells die for organisms to grow. Ecosystems require decomposition to recycle nutrients. What we call death is transformation—matter and energy changing forms while serving larger purposes. Dinosaur fossils became petroleum, fueling human civilization. Their death served life.

Some theistic evolutionists go further, seeing death as part of creation's freedom. God granted creation genuine autonomy, including possibilities for suffering. This makes love, courage, and growth possible. A world without challenge would be static. Dinosaur predators and prey drove evolutionary innovation, creating the dynamic biosphere we inhabit.

The Problem of Divine Action

How Does God Guide Evolution?

If God works through evolution, how does He ensure specific outcomes? The process appears so contingent—asteroids might miss, mutations might differ, selections might vary. Stephen Jay Gould famously argued that "rewinding life's tape" would produce completely different results. How can such a process reliably produce God's intentions?

Theistic evolutionists propose several models of divine action. Some embrace "front-loading"—God designed initial conditions and laws to inevitably produce desired outcomes. The constants of physics, the properties of carbon, the logic of natural selection all conspire toward life's emergence and development. Dinosaurs arose not through miraculous intervention but because creation's fabric was woven to produce them.

Others propose ongoing divine influence at quantum levels. Quantum events are genuinely indeterminate—God could influence outcomes without violating natural laws. Accumulated over billions of years, tiny nudges could guide evolution precisely. This preserves both scientific descriptions and theological convictions about providence.

Still others speak of "convergent creation"—God's purposes achieved through multiple pathways. If one lineage fails, others fulfill the role. If dinosaurs hadn't evolved, other creatures would have filled their niches. The specific forms matter less than functions. God's intentions are robust, achieved through various means.

The Scandal of Particularity

Why did God use such a slow, meandering process? Why allow dinosaurs to rule Earth for 165 million years before creating humans? The "scandal of particularity"—God working through specific, limited means—appears throughout Scripture. He chose Abraham's family, not all humanity. He revealed Himself to Israel, not every nation. The Incarnation occurred once in history, not repeatedly.

Similarly, God chose evolution as His creative method despite apparent inefficiencies. The process that produced dinosaurs also generated countless "failed" species, dead-end lineages, and extinctions. Yet Scripture reveals God delighting in creation's diversity. Every sparrow matters; how much more every Compsognathus?

The timescales involved—billions of years—challenge human comprehension but not divine patience. Psalm 90 reminds us that God perceives time differently. What seems slow to us may be swift to Him. The age of dinosaurs lasted 165 million years, but this is merely 3% of Earth's history, a blink in cosmic time.

God crafts masterpieces without hurry.

Theological Richness in Deep Time

Creation's Freedom and Creativity

Theistic evolution emphasizes creation's genuine freedom. Rather than micromanaging every detail, God granted creation capacities for development. Natural laws aren't restrictions but enablements—allowing matter to organize, life to emerge, and complexity to increase. Dinosaurs exemplify this creative freedom.

Consider feathered dinosaurs discovered in China. Each species displays unique variations—some with simple filaments, others with complex plumes. This diversity reflects not divine indecision but creation's exploratory freedom. God delights in novelty emerging from processes He established. Every new dinosaur species revealed creation's potential.

This theological vision connects with biblical themes of growth and maturation. Jesus's parables emphasize gradual development—seeds growing secretly, dough rising slowly, mustard seeds becoming trees. Creation follows similar patterns. The universe itself matures, moving from simplicity toward complexity, from potential toward actualization.

The Cosmic Christ and Deep History

Some theistic evolutionists draw on biblical theology about Christ's cosmic role. Colossians 1:16-17 declares all things created through and for Christ, who holds everything together. John 1 identifies Christ as the Logos through whom everything was made. This cosmic Christ works throughout creation's history, including the age of dinosaurs.

From this perspective, evolution itself reveals Christ's character—patient, creative, suffering with creation while guiding it toward glory. The long history of life, with its struggles and extinctions, parallels the paschal mystery. Death and resurrection patterns appear throughout nature. Dinosaur extinction enabled mammalian flourishing, prefiguring how Christ's death brings new life.

This doesn't make dinosaurs "Christian" but locates them within salvation history broadly conceived. All creation groans awaiting redemption (Romans 8:22). Dinosaurs participated in this cosmic drama, their fossils bearing witness to creation's bondage to decay and hope for transformation.

Pastoral and Practical Implications

Science as Doxology

For theistic evolutionists, studying dinosaurs becomes worshipful. Each fossil reveals God's creativity. Each evolutionary innovation demonstrates divine wisdom. Rather than threatening faith, paleontology deepens wonder. Young Christians can pursue scientific careers as vocations, exploring God's world to glorify Him.

This approach transforms science education. Rather than defensive antagonism, Christians can enthusiastically engage evolutionary biology. Yes, some presentations assume atheistic materialism, but the science itself remains neutral. Students can learn evolutionary theory while maintaining vibrant faith, seeing God's hand throughout.

Churches embracing theistic evolution often celebrate science. They host creation care initiatives, support members in scientific careers, and integrate scientific insights into worship. Dinosaur exhibits become opportunities for praising God's ancient works rather than battlegrounds for apologetics.

Humility and Mystery

Theistic evolution cultivates intellectual humility. We're discovering how God created, not prescribing methods He must use. Our theological frameworks must remain flexible enough to incorporate new discoveries. What we learn about dinosaurs tomorrow might revise today's understandings.

This humility extends to biblical interpretation. Genesis teaches crucial theology—God as sole Creator, creation's goodness, human uniqueness—without providing scientific details. Ancient Israelites didn't need to know about dinosaurs to understand their relationship with God. We can hold firm to theological truths while remaining open about scientific particulars.

Mystery permeates this approach. How exactly does God guide evolution? When precisely did humans receive God's image? Why did God choose this particular method? Theistic evolutionists often answer, "We don't fully know." This isn't evasion but honest recognition of our limitations before infinite God.

Criticisms and Responses

From Young-Earth Creationists

Young-earth advocates charge theistic evolution with compromising biblical authority. If Genesis 1 doesn't mean what it plainly says, how do we know when to take Scripture literally? Death before sin undermines the gospel. Millions of years of suffering contradict God's goodness.

Theistic evolutionists respond that they take Scripture seriously, not literally where genre indicates otherwise. Genesis 1 uses ancient cosmology (waters above and below) we don't accept scientifically. The text's theological truths transcend its cultural packaging. Regarding death, they distinguish human spiritual death from animal mortality, preserving the gospel while accepting paleontological evidence.

From Intelligent Design Proponents

ID theorists argue theistic evolution grants too much to naturalism. If evolution explains everything, where's evidence for God? The position seems indistinguishable from atheistic evolution with God tacked on. Real design requires detectable divine intervention, not just evolutionary processes.

Theistic evolutionists counter that God's action needn't be scientifically detectable to be real. We don't see God's hand directly in sunrise, yet He sustains it. Natural processes themselves evidence design—why should matter organize into increasingly complex forms? The elegance of evolution testifies to divine wisdom more than sporadic interventions would.

From Atheistic Evolutionists

Secular scientists often dismiss theistic evolution as unnecessary. If evolution works without God, why invoke Him? Occam's Razor favors purely natural explanations. Adding God to evolution is like adding fairies to gravity—superfluous and unscientific.

Theistic evolutionists respond that science addresses mechanisms, not ultimate causes. Evolution describes how species change but not why the universe exists, why laws permit evolution, or why consciousness emerged. These philosophical and theological questions transcend science. God isn't added to evolution but recognized as its ground and goal.

The Future of Faith and Fossils

Theistic evolution continues developing sophistication. Scholars explore how divine action might work, how human uniqueness emerged, and how theology engages science. Dinosaur discoveries particularly excite this community—each fossil reveals more of God's creative process.

The approach's ecumenical potential deserves note. Catholics, mainline Protestants, and increasing numbers of evangelicals find common ground here. While disagreeing on biblical interpretation or church authority, they unite in seeing evolution as God's method. This shared perspective enables collaborative witness in scientifically educated cultures.

For dinosaur enthusiasts, theistic evolution offers rich possibilities. Every museum visit becomes pilgrimage. Each new species discovered reveals divine creativity. The entire Mesozoic Era testifies to God's patient artistry. Rather than problems requiring solution, dinosaurs become windows into divine action across deep time.

Looking Ahead

Our next chapter examines one of the most challenging aspects of dinosaur theology—the problem of natural evil. How do we reconcile T. rex's terrible teeth with divine benevolence? What does predation say about God's character? These questions press whether we accept young-earth, old-earth, or evolutionary perspectives.

The dinosaur fossil record confronts us with death, violence, and suffering long before human sin. Mass extinctions, evidence of disease, and clear predation challenge simple theodicies. Yet Christians throughout history have found ways to affirm both creation's goodness and its genuine travails. Their wisdom, combined with modern insights, may help us understand why God's "very good" world included terrible lizards.

For Reflection

1. Does viewing evolution as God's creative method enhance or diminish your sense of divine majesty?
2. How do you understand the relationship between natural processes and divine action?
3. What aspects of theistic evolution do you find most compelling? Most troubling?

Did You Know?

Pope John Paul II declared in 1996 that evolution is "more than a hypothesis," accepting it as established science while maintaining God's special creation of human souls. Pope Francis continues this tradition, stating that God is not "a magician with a magic wand" but works through natural processes. The Catholic Church's acceptance of evolution while maintaining theological orthodoxy demonstrates that biblical faith and evolutionary science can coexist.

Consider This

The only dinosaurs that survived the mass extinction were birds–small, feathered theropods that found new niches when their larger relatives perished. From a theistic evolutionary perspective, this transformation from terrible lizards to singing sparrows exemplifies creation's resurrection pattern. What seemed like ending became new beginning, death became life, scales became feathers fit for flight.

For Further Study

* Collins, Francis. The Language of God
* Haarsma, Deborah and Loren. Origins: Christian Perspectives on Creation, Evolution, and Intelligent Design
* McGrath, Alister. A Fine-Tuned Universe
* Lamoureux, Denis. Evolutionary Creation
* BioLogos.org for extensive resources on evolutionary creation

Chapter 8:

Tyrannosaurus and the Problem of Evil

"The young lions roar for their prey, seeking their food from God." — Psalm 104:21

The Killing Machine

Sue stands thirteen feet tall at the hips, stretching forty feet from nose to tail. Her skull alone weighs 600 pounds, housing sixty teeth designed for one purpose: crushing bone. The largest teeth, protruding eight inches from the jaw, are thick as railroad spikes and serrated like steak knives. Her bite force—12,800 pounds per square inch—exceeds any land animal living or extinct. She is Tyrannosaurus rex, the tyrant lizard king, and she poses one of theology's hardest questions.

Sue (named after discoverer Sue Hendrickson) resides at Chicago's Field Museum, where millions marvel at her terrible magnificence. Children press against the glass, simultaneously thrilled and terrified. But for thoughtful Christians, Sue represents more than prehistoric spectacle. She embodies the problem of natural evil—how could a good God create such an engine of destruction?

The evidence is unambiguous. Sue's bones show healed injuries from violent encounters. Her coprolites (fossilized feces) contain crushed bone fragments from prey. Most tellingly, we've found Triceratops bones bearing T. rex bite marks that partially healed, proving the attacks occurred while victims lived. This wasn't scavenging but active predation. Sue killed to live, and she was very, very good at it.

Design for Death

The Anatomy of a Predator

Every aspect of Tyrannosaurus rex seems optimized for predation. The massive skull supported powerful jaw muscles anchored to enlarged cranial crests. The teeth weren't just sharp but precisely engineered—front teeth gripped, side teeth sliced, back teeth crushed. They continuously replaced throughout life, ensuring fresh weapons always ready.

The eyes faced forward, providing binocular vision for depth perception crucial in striking prey. CAT scans reveal enlarged olfactory bulbs, indicating exceptional smell for tracking wounded animals. Even the tiny arms, often mocked, were incredibly strong—each could curl 400 pounds—possibly used to grip struggling prey while teeth delivered the killing bite.

But T. rex wasn't uniquely terrible. Spinosaurus, possibly larger, possessed crocodile-like jaws perfect for catching fish—though "fish" included sharks and car-sized lungfish. Utahraptor deployed switchblade claws to disembowel prey. Allosaurus hunted in packs, revealed by synchronized tooth marks on sauropod bones. The Mesozoic was an arms race of offensive and defensive weaponry.

This wasn't limited to dinosaurs. The seas hosted nightmares like Mosasaurus, with jaws that unhinged like a snake's to swallow prey whole. Pteranodon snatched fish from above with spear-like beaks. Even herbivorous dinosaurs inflicted violence—Triceratops horns could gore predators, Ankylosaurus tail clubs shattered bones, Therizinosaurus wielded three-foot claws.

The Suffering of the Mesozoic

The fossil record preserves moments of ancient agony frozen in stone. The famous "Fighting Dinosaurs" specimen captures a Velociraptor with its sickle claw embedded in a Protoceratops's neck, while the herbivore's beak crushes the predator's arm. Both

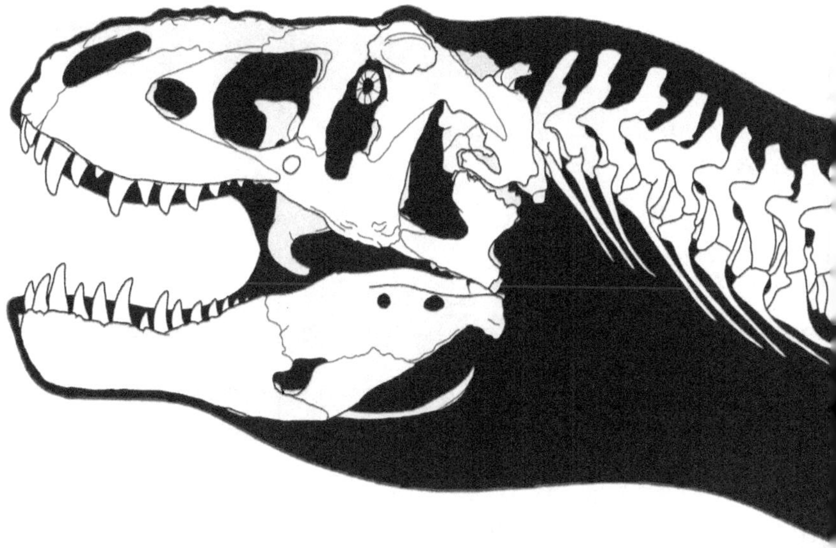

died in mortal combat, buried by a sudden sandstorm.

Disease plagued dinosaurs too. We find evidence of bone cancer, arthritis, gout, and infections. Parasites left traces in coprolites. Healed fractures show survival through painful recovery. One Dilophosaurus specimen reveals a deformed humerus that would have caused lifelong agony. An Allosaurus named "Big Al" showed nineteen different pathologies—a lifetime of accumulated suffering.

Even birth was dangerous. Fossil eggs sometimes contain embryos that died before hatching. Nesting sites reveal high infant mortality. For every T. rex that reached adulthood, dozens died young—starved, diseased, or devoured. Nature was "red in tooth and claw" long before Tennyson coined the phrase.

Theological Wrestling

The Classic Dilemma

The problem of evil has challenged believers since Job, but dinosaurs sharpen it uniquely. Human suffering might result from

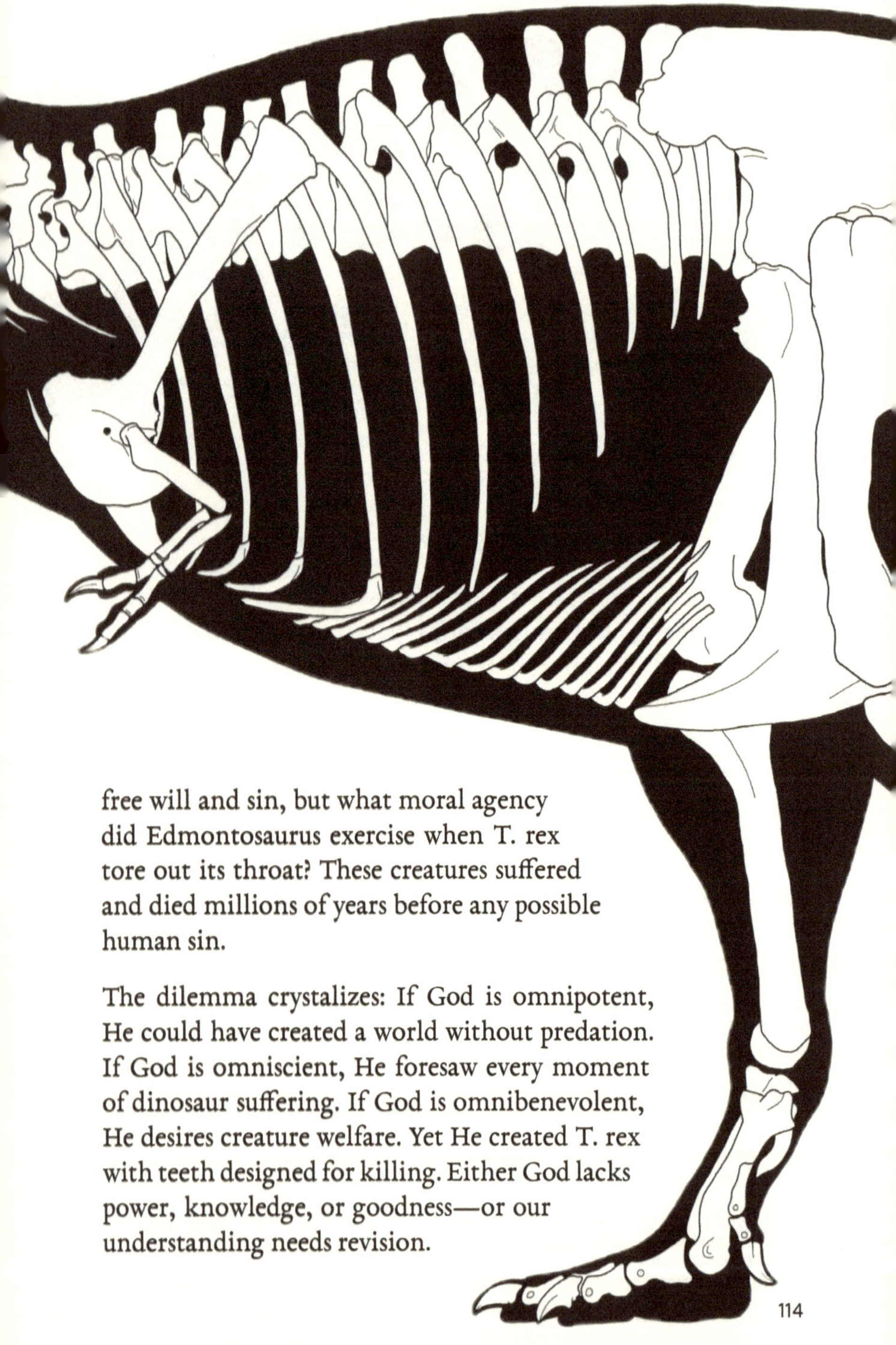

free will and sin, but what moral agency did Edmontosaurus exercise when T. rex tore out its throat? These creatures suffered and died millions of years before any possible human sin.

The dilemma crystalizes: If God is omnipotent, He could have created a world without predation. If God is omniscient, He foresaw every moment of dinosaur suffering. If God is omnibenevolent, He desires creature welfare. Yet He created T. rex with teeth designed for killing. Either God lacks power, knowledge, or goodness—or our understanding needs revision.

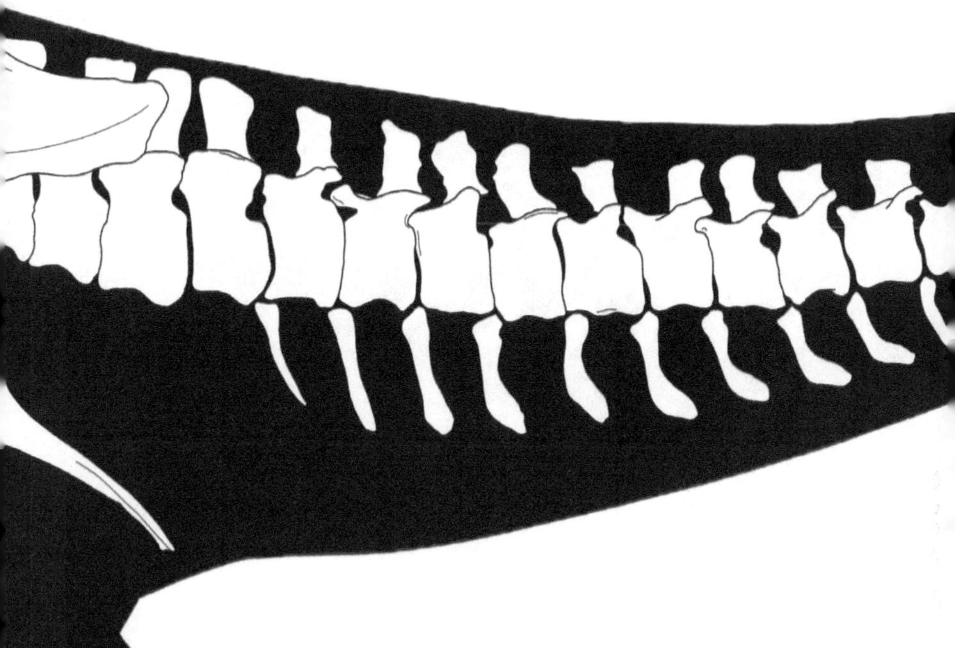

Young-earth creationists escape by denying pre-human suffering. All dinosaur death occurred after Adam's sin, they argue. But this requires rejecting mainstream geology, paleontology, and physics. It also raises questions about why sin would transform herbivore teeth into carnivore fangs. Did God redesign creation for cruelty as punishment?

Old-Earth Responses

Christians accepting geological time must confront millions of years of pre-human suffering. Several theological strategies have emerged, each with strengths and weaknesses.

The Free Process Defense: Just as God

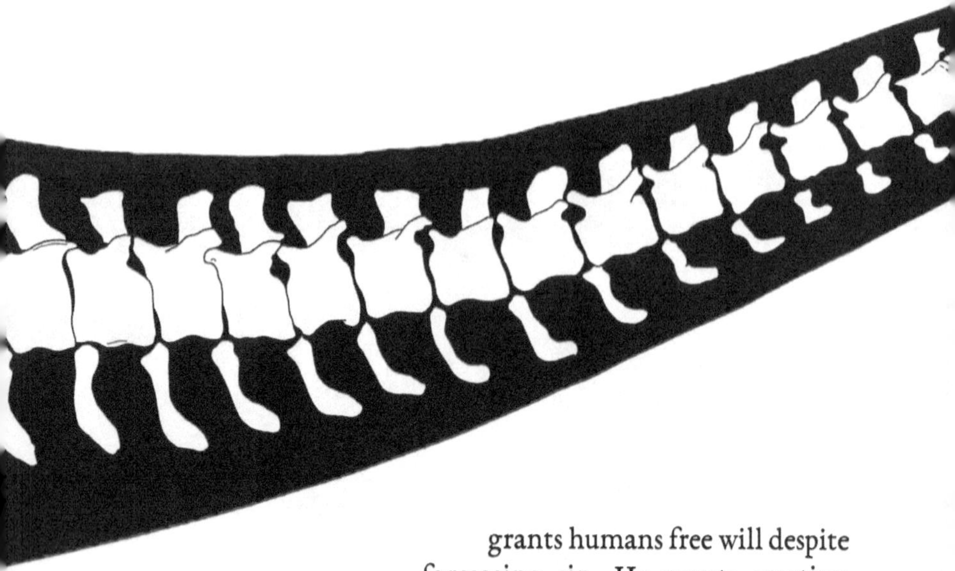

grants humans free will despite foreseeing sin, He grants creation freedom to develop naturally. Evolution requires selection pressure—without predation, there's no driver for adaptation. T. rex represents creation's freedom working through natural laws. God permits suffering for greater goods: diversity, beauty, and ultimately consciousness capable of relationship with Him.

Critics respond that comparing natural processes to moral agents seems strained. Humans choose sin; did prehistoric seas choose to evolve Mosasaurus? This defense also struggles with the intensity of suffering. Couldn't God have designed evolution to work through less violent means?

The Greater Good Theodicy: Perhaps predation serves purposes we don't fully grasp. Predators prevent overpopulation and starvation. They strengthen prey species by eliminating the weak. The fear they inspire drives evolutionary innovation—better defenses, keener senses, social cooperation. Without T. rex, would Triceratops have developed its magnificent frill and horns?

This utilitarian calculus troubles many. Are we really arguing

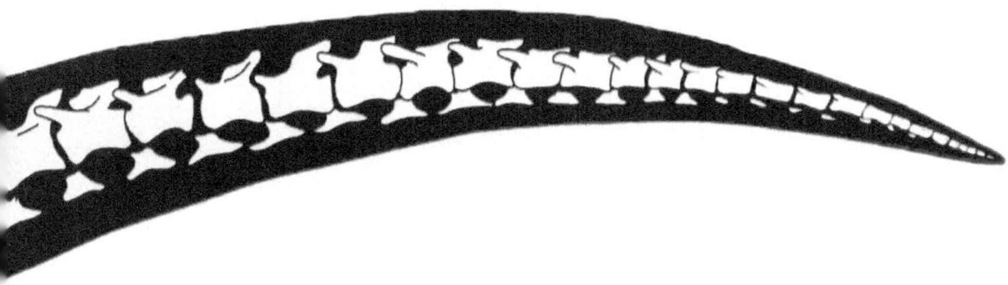

that baby Maiasaura needed to die horribly for evolution to produce better dinosaurs? The goods achieved seem incommensurate with suffering endured. Would a truly omnipotent God require such brutal means for biological ends?

The Aesthetic Defense: Some theologians propose that God values drama, intensity, and even tragedy in creation's story. T. rex adds narrative power to natural history. The interplay of predator and prey creates dynamic ecosystems more interesting than static peace. Like a novelist including conflict, God crafts a creation with genuine stakes.

This aestheticizing of suffering strikes many as morally abhorrent. Did Parasaurolophus die in Daspletosaurus jaws for divine entertainment? The aesthetic defense works better for human art, where suffering is represented, not real. Actual creatures experienced actual agony.

Natural Evil Versus Moral Evil

A crucial distinction in theodicy separates natural from moral evil. Moral evil results from free agents choosing wrongly—murder, theft, cruelty. Natural evil involves suffering without moral agency—earthquakes, diseases, predation. Dinosaur suffering was entirely natural evil.

Some argue natural evil isn't truly evil—it's just nature functioning. T. rex killing Edmontosaurus is no more evil than lightning striking trees. These are physical processes following natural laws. Evil requires intention, which natural processes lack. Calling predation "evil" projects human categories onto amoral nature.

Yet this seems to diminish real suffering. The Edmontosaurus experienced terror and agony regardless of T. rex's moral status. Pain is intrinsically bad for the creature experiencing it. That millions of sentient beings suffered for millions of years before humans existed remains troubling, whatever categories we apply.

Death in "Very Good" Creation

Reexamining Genesis 1:31

After creating everything, "God saw all that he had made, and it was very good." How do we reconcile this pronouncement with a world containing T. rex? Several interpretive strategies emerge.

Some argue "very good" means "fit for purpose" rather than "morally perfect." Creation was very good for accomplishing God's intentions—developing complexity, beauty, and ultimately beings capable of bearing His image. T. rex was very good at being T. rex, fulfilling its ecological role perfectly.

Others suggest "very good" describes creation's potential rather than every present reality. The world was very good as a launching point for development. Just as human infants are "very good" despite inability and dependency, creation was "very good" despite containing processes involving suffering.

Still others interpret "very good" relationally—creation perfectly expressed God's intentions at that moment. As a master chef declares each cooking stage "very good" despite raw ingredients or intermediate steps humans wouldn't eat, God pronounced each creative stage good for what it was, not as final product.

The Christological Lens

Some theologians find help in Christology. The same God who created T. rex entered creation in Christ, subjecting Himself to suffering and death. The Cross reveals a God who doesn't observe suffering distantly but enters it personally. Divine solidarity with suffering creation changes how we view prehistoric pain.

Furthermore, Christ's resurrection promises creation's ultimate redemption. Romans 8 describes creation groaning in bondage, awaiting liberation. Perhaps predation and death were always intended as temporary, preparing for transformation. T. rex and its victims participate in the cosmic drama culminating in new creation.

This doesn't explain why God chose this particular path involving so much suffering. But it assures us that the God who allowed Velociraptor to slash Protoceratops also experienced claws and thorns Himself. Divine empathy doesn't solve the problem of evil but prevents us from imagining an indifferent deity.

Insights from Predation

The Ecological Perspective

Modern ecology reveals predation's crucial role in healthy ecosystems. The reintroduction of wolves to Yellowstone demonstrates this dramatically. Without predators, deer overpopulated, devastating vegetation. Rivers eroded as plants disappeared. Biodiversity plummeted. Wolves' return restored balance, indirectly benefiting species from beavers to butterflies.

Similarly, Mesozoic ecosystems required apex predators like T. rex. Computer models suggest that without large theropods, herbivore populations would have exploded, stripping vegetation and causing ecological collapse. Predation, however brutal individually, maintained systemic health. The tragedy of individual death enabled collective flourishing.

This ecological perspective doesn't eliminate the problem but reframes it. Death serves life. Suffering maintains balance. What appears cruel in isolation proves necessary in context. God's creation includes interconnected systems where individual goods sometimes conflict with collective goods. The gazelle's perspective differs from the ecosystem's.

The Evolutionary Arms Race

Predation drove remarkable innovations. Ankylosaurus armor plating, Pachycephalosaurus skull domes, and Styracosaurus horn arrays all evolved in response to predator pressure. Similarly, predators evolved greater speed, sharper senses, and better weapons to overcome defenses. This "arms race" produced the spectacular diversity we marvel at in museums.

Without predation, evolution stagnates. Why develop better eyesight without threats to see? Why grow faster without pursuers? Why cooperate without shared danger? The very features we find beautiful—elaborate frills, colorful displays, complex behaviors—often originated as survival adaptations against predators.

From this perspective, T. rex was an unwitting agent of creation, driving other species toward greater complexity and capability. The suffering it caused catalyzed developments that wouldn't otherwise occur. This doesn't justify suffering but suggests it served purposes beyond immediate destruction.

Modern Parallels and Ancient Wisdom

Nature's Continuing Violence

Dinosaurs didn't invent predation—they inherited and elaborated it. Predation existed in the seas before land colonization. It continues today in forms equally brutal if usually smaller scale. The same theological questions raised by T. rex apply to lions, sharks, and parasitic wasps.

Darwin famously struggled with ichneumon wasps that lay eggs in living caterpillars, which are slowly devoured from within. He wrote, "I cannot persuade myself that a beneficent and omnipotent God would have designedly created the Ichneumonidae with the express intention of their feeding within the living bodies of Caterpillars."

Yet Scripture seems remarkably comfortable with predation. Psalm 104 celebrates lions seeking prey from God. Job 38-39 glories in wild animals' freedom, including predatory birds. Jesus used wolves, serpents, and birds of prey in teachings without condemning their nature. Biblical writers saw predation as part of creation's fabric, not evidence against God's goodness.

The Limits of Understanding

The book of Job provides the most extensive biblical treatment of suffering and divine justice. Significantly, God's answer to Job involves extensive nature poetry, including possible dinosaur references (Behemoth and Leviathan). Rather than explaining suffering philosophically, God points to creation's wildness, mystery, and magnificence.

The message seems to be that creation exceeds human comprehension. We lack the perspective to judge whether a world with T. rex is better than imaginable alternatives. Our finite minds can't grasp infinite wisdom's calculations. This isn't evasion but honest acknowledgment of creaturely limitations.

Modern chaos theory reinforces this humility. In complex systems, tiny changes cascade into massive differences. Removing T. rex from the Cretaceous might have prevented human evolution through butterfly effects we can't calculate. What seems cruel might be necessary for goods we can't foresee.

Living with the Mystery

Pastoral Perspectives

How do we help struggling believers reconcile T. rex with divine love? Several pastoral approaches prove helpful:

First, acknowledge the genuine difficulty. Don't minimize prehistoric suffering or offer glib solutions. The fossil record documents real agony that troubles thoughtful believers. Validating these concerns opens honest dialogue.

Second, explore multiple perspectives without demanding single answers. Present various theodicies as partial insights rather than complete solutions. Different approaches resonate with different people. Some find ecological explanations helpful; others prefer Christological responses.

Third, emphasize what we know about God through Christ over what we don't understand about creation. The Cross demonstrates divine love more clearly than fossils demonstrate divine indifference. Ground faith in revelation's certainties rather than nature's ambiguities.

Finally, cultivate wonder alongside wrestling. T. rex raises hard questions but also inspires awe. The same God who allowed its terrible reign also crafted its magnificent form. Mystery encompasses both beauty and terror in creation's drama.

The Continuing Quest

The problem of natural evil remains theology's hardest challenge. Dinosaurs make it unavoidable—millions of years of suffering before any possible Fall. Yet believers throughout history have maintained faith despite nature's cruelty. They've found ways to affirm both creation's goodness and its genuine travails.

Perhaps the answer isn't a single solution but a symphony of partial insights. Ecological necessity, evolutionary creativity, divine solidarity, eschatological hope, and creaturely limitations all contribute perspectives. Together they don't solve the problem but make it bearable.

T. rex forces us to wrestle with God like Jacob, refusing easy answers. This wrestling, painful as it is, deepens faith. We discover a God complex enough to create both terrible lizards and suffering servants, both Cretaceous carnage and Calvary's love. The mystery remains, but so does the faith.

Looking Forward

Our next chapter examines another challenge dinosaurs pose to faith—the question of time. How do we reconcile biblical chronology with geological ages? Can Earth be both thousands and billions of years old? The debate over dating methods and deep time proves nearly as contentious as theodicy.

Yet as we'll see, the question of when dinosaurs lived intertwines with questions of how we know, what we trust, and where we place our ultimate confidence. Time, like suffering, pushes us beyond easy answers toward deeper understanding of both God's word and God's world.

For Reflection

1. How do you personally reconcile the existence of predatory dinosaurs with belief in a good Creator?
2. Which theological response to natural evil do you find most helpful? Most troubling?
3. Does learning about prehistoric suffering challenge or deepen your faith? Why?

Did You Know?

Dracorex hogwartsia wasn't the only dinosaur with a whimsical name. Bambiraptor was named for Disney's deer due to its small size and large eyes. Yet this "Bambi" was a predator with razor-sharp claws and teeth. Even our naming conventions reveal discomfort with nature's violence–we project innocence onto efficient killers.

Consider This

Some T. rex specimens show evidence of cannibalism–tooth marks from their own species on bones. This ultimate predator sometimes turned on its own kind. From one perspective, this represents nature's ultimate cruelty. From another, it shows even apex predators faced mortality's limits. The tyrant king was subject to the same brutal rules governing all life.

For Further Study

* Southgate, Christopher. The Groaning of Creation: God, Evolution, and the Problem of Evil
* Murray, Michael. Nature Red in Tooth and Claw: Theism and the Problem of Animal Suffering
* Sollereder, Bethany. God, Evolution, and Animal Suffering
* Polkinghorne, John, ed. The Work of Love: Creation as Kenosis
* Lewis, C.S. The Problem of Pain (chapter on animal suffering)

Chapter 9:

The Time of Dinosaurs: Millions of Years or Thousands?

"But do not overlook this one fact, beloved, that with the Lord one day is as a thousand years, and a thousand years as one day." — 2 Peter 3:8

The Clock in the Rocks

Dr. Andrew Snelling stood before the Grand Canyon's towering walls, seeing two completely different stories written in stone. As a PhD geologist who worked for years in the mining industry, he understood conventional geology's reading: these layers accumulated over hundreds of millions of years, preserving Earth's deep history. But as a committed young-earth creationist, he saw evidence of Noah's flood—massive deposits laid down rapidly in a global catastrophe just thousands of years ago.

This isn't merely an academic dispute. The age of Earth's rocks determines when dinosaurs lived, which shapes how we understand their relationship to humans and their place in biblical history. If radiometric dating is reliable, dinosaurs died out 66 million years before humans appeared. If Earth is young, humans and dinosaurs were contemporaries. The stakes couldn't be higher for understanding God's creation.

The battle over time has divided Christians for two centuries. Families split over the issue. Denominations exclude those who disagree. Scientists who question conventional dating risk career suicide. Yet the questions persist: How do we really know the age of rocks? What assumptions underlie dating methods?

Can both Scripture and science be trusted when they seem to conflict about time?

How Dating Works: Promises and Perils

The Radiometric Revolution

The discovery of radioactivity in 1896 revolutionized geology. Scientists realized that unstable elements decay at constant rates, providing potential "clocks" for dating rocks. By measuring the ratio of parent isotopes (original radioactive element) to daughter products (decay results), researchers could calculate how long decay had been occurring.

Different radioactive elements decay at different rates, measured in half-lives—the time required for half the parent isotope to decay. Carbon-14 has a half-life of 5,730 years, useful for recent materials. Potassium-40's half-life of 1.25 billion years dates older rocks. Uranium-238, with a 4.5-billion-year half-life, can date Earth's oldest minerals.

For dinosaur fossils, several methods apply. The bones themselves can't be directly dated—fossilization replaces original material with minerals. Instead, scientists date volcanic ash layers above and below fossil-bearing sediments. If ash below dinosaur bones dates to 70 million years and ash above dates to 65 million years, the dinosaurs lived between those times.

Multiple dating methods often cross-check each other. The K-T boundary (now called K-Pg boundary) marking dinosaur extinction has been dated by potassium-argon, argon-argon, uranium-lead, and rubidium-strontium methods at numerous sites worldwide. All converge on approximately 66 million years ago. This consistency across methods and locations builds confidence in the results.

Critical Assumptions

However, radiometric dating requires several assumptions that young-earth creationists challenge:

Initial Conditions: We must know the starting ratio of parent to daughter isotopes. Conventional geologists argue that daughter products can be identified and excluded, but critics question whether we can truly know conditions millions of years ago. Contamination or inheritance of daughter isotopes would yield falsely old ages.

Closed System: The rock must remain closed to addition or removal of parent or daughter isotopes except by radioactive decay. But rocks experience weathering, metamorphism, and groundwater infiltration. Critics argue these processes could alter isotope ratios, making ages unreliable. Defenders counter that careful sample selection and multiple analyses can detect and avoid altered rocks.

Constant Decay Rates: Perhaps most controversially, radiometric dating assumes decay rates never change. This seems reasonable—extensive testing shows decay rates unaffected by temperature, pressure, or chemical environment. But young-earth creationists propose past conditions might have accelerated decay dramatically.

The RATE project (Radioisotopes and the Age of The Earth), conducted by creation scientists, investigated whether decay rates might have been millions of times faster in the past. They point to several lines of evidence: helium retention in zircons suggesting accelerated alpha decay, radiohalos indicating rapid uranium decay, and discordant isochron dates possibly reflecting non-constant decay.

The Carbon-14 Conundrum

One of young-earth creationism's strongest arguments involves carbon-14 in supposedly ancient materials. C-14's short half-life means it should be undetectable after about 100,000 years—all would have decayed away. Yet researchers report finding C-14 in coal, diamonds, and even dinosaur bones allegedly millions of years old.

Dr. John Baumgardner, a geophysicist with a PhD from UCLA, has championed this research. His team sent dinosaur bone samples to commercial dating laboratories without revealing their source. Results consistently showed detectable C-14, suggesting ages of thousands, not millions of years. How can 70-million-year-old T. rex bones contain any C-14?

Conventional scientists propose contamination—modern carbon infiltrating ancient samples through groundwater, microbes, or laboratory handling. They note that the detected C-14 levels are extremely low, consistent with contamination rather than original carbon. But young-earth researchers argue they've taken extensive precautions against contamination, and the consistency across many samples suggests genuine endogenous carbon.

The Geological Column: Order from Chaos?

Reading the Rocks

Beyond radiometric dating, the sequence of rock layers tells a story. The geological column—the idealized sequence of rock layers containing distinctive fossils—appears remarkably consistent worldwide. Cambrian rocks contain trilobites but no dinosaurs. Jurassic rocks yield Stegosaurus but no flowers. Cretaceous layers have T. rex and flowering plants but no humans.

Conventional geology interprets this sequence as temporal—lower layers are older, containing earlier life forms. The order makes evolutionary sense: marine invertebrates, then fish, then land plants, amphibians, reptiles, dinosaurs, mammals, and finally humans. Each layer represents millions of years of deposition and evolution.

Young-earth geologists see the same rocks but interpret them differently. The order, they argue, reflects not time but flood dynamics. During Noah's flood, marine creatures were buried first, then lowland animals, then upland creatures. Hydrological sorting by size, density, and mobility created the apparent sequence. Smarter, more mobile humans avoided burial until post-flood local catastrophes.

This flood geology faces challenges. Why don't we find any flowering plants in Carboniferous coal deposits, even though modern floods mix vegetation indiscriminately? Why are microscopic foraminifera so precisely sorted into zones useful for oil exploration? How did footprints and burrows form in the middle of a violent flood? The ordered sequence seems to reflect more than hydraulic sorting.

Rapid Formation Features

Yet young-earth geologists point to features suggesting rapid deposition. Polystrate fossils—usually tree trunks extending through multiple rock layers—challenge slow accumulation. How did trees remain upright for millions of years while sediment slowly buried them? Conventional geology invokes rapid local events within overall slow deposition, but critics see evidence for rapid formation of entire sequences.

The Mount St. Helens eruption in 1980 provided a natural laboratory for catastrophic geology. In hours and days, volcanic flows deposited layered sediments resembling formations supposedly requiring millions of years. Mudflows carved canyons through solid rock. If local catastrophes produce such features rapidly, what could a global flood accomplish?

Soft sediment deformation also suggests rapid deposition. Many rock layers show folding and bending while still soft, before lithification. The Grand Canyon's Tapeats Sandstone and Redwall Limestone show folding without fracturing, implying the entire sequence was still soft when deformed. This seems to require deposition and deformation within a short timeframe, not millions of years.

Alternative Chronologies

The Appearance of Age

Some Christians propose that God created the universe with apparent age. Just as Adam was created mature, not infant, perhaps God created a mature universe. Light from distant stars was created in transit. Rocks were created with appropriate isotope ratios. Fossils were created in rocks to provide resources and reveal God's creativity.

This "omphalos" hypothesis (named after the Greek word for navel—did Adam have one?) solves chronological problems but creates theological ones. It makes God deceptive, creating false history. If fossils don't represent real creatures that lived and died, what do they mean? Most reject this solution as incompatible with God's truthfulness.

A variation suggests functional maturity without false history. Wine at Cana was created with properties of aged wine but without false years in non-existent barrels. Similarly, creation might possess functional maturity—light from stars visible immediately, ecosystems fully operational—without deceptive history. But fossils of dead creatures seem to record actual events, not just functional features.

Time Dilation Models

More sophisticated young-earth models invoke relativistic physics. Dr. Russell Humphreys proposes that gravitational time dilation during creation week allowed billions of years to pass in distant space while only days passed on Earth. Einstein's relativity shows time isn't absolute—it varies with gravity and velocity. Near the massive gravity of early creation, Earth time slowed dramatically.

This model allows distant starlight to reach Earth within a young-earth timeframe without invoking appearance of age. It accepts standard physics and astronomy for the distant universe while maintaining a young Earth. Critics question whether the required gravitational fields match astronomical observations and whether the model explains geological features on Earth itself.

Dr. Jason Lisle's Anisotropic Synchrony Convention takes a different approach. He notes that the one-way speed of light is conventionally defined, not measured. By choosing a different synchrony convention—instantaneous light travel toward observers—distant starlight arrives immediately. This solution is mathematically valid within relativity but conceptually challenging and doesn't address geological aging.

Testing the Tests

Discordant Dates

Both sides cite cases where radiometric dating yields problematic results. Young-earth creationists compile lists of discordant dates—rocks of known age dating incorrectly, different methods giving different ages, or clearly related rocks yielding vastly different dates. Mount St. Helens' 1980 lava dome dated to 350,000 years by potassium-argon dating, clearly wrong for a 10-year-old rock.

Conventional geologists respond that such examples usually involve inappropriate sample selection or dating methods. The Mount St. Helens dacite contained xenoliths—older rock fragments incorporated into the lava—that contaminated the sample. Potassium-argon dating assumes no initial argon, but volcanic rocks can trap atmospheric argon. Proper protocols avoid such errors.

The question becomes whether occasional errors invalidate the entire enterprise or whether successful dates far outnumber failures. Oil companies rely on radiometric dates for exploration. If the methods were fundamentally flawed, why would profit-driven corporations use them? Young-earth advocates respond that relative dating works for finding oil without requiring absolute ages to be correct.

Soft Tissue Survival

Mary Schweitzer's discovery of soft tissue in dinosaur fossils shocked the scientific community. Proteins should decay relatively rapidly—laboratory experiments suggest complete degradation within thousands to millions of years under ideal conditions. How can organic molecules survive 70 million years?

Initial skepticism was intense. Could these be biofilms mimicking blood vessels? Contamination from modern sources? But follow-up studies confirmed original proteins, including collagen, elastin, and possibly even DNA fragments. The tissues show remarkable preservation, including apparent red blood cells and bone cells with nuclei.

Young-earth creationists see this as powerful evidence for recent burial. If dinosaurs died thousands, not millions of years ago, soft tissue preservation becomes expected rather than miraculous. Conventional scientists propose iron from blood acts as a preservative, cross-linking proteins and protecting them from decay. Experiments show iron can extend protein survival, though whether it can preserve tissues for tens of millions of years remains debated.

Living with Uncertainty

The Humility of Not Knowing

The time question reveals the limits of both scientific and theological knowledge. We weren't there to observe dinosaurs living or dying. We can't directly measure past decay rates or verify initial conditions. Our interpretations of both rocks and Scripture involve assumptions that might be wrong.

This uncertainty disturbs those seeking absolute answers but can foster healthy humility. Perhaps God allows ambiguity to prevent idolatry of either science or particular biblical interpretations. The young-earth believer must honestly acknowledge scientific challenges to recent creation. The old-earth believer must recognize that conventional dating involves unprovable assumptions.

Rather than weaponizing uncertainty—"You can't prove the Earth is old!"—we might embrace it as invitation to deeper investigation. Each challenge spurs better understanding. Young-earth research has improved radiometric protocols. Conventional responses have clarified dating methods' logic and limitations. Iron sharpens iron through respectful disagreement.

Core Convictions and Peripheral Perspectives

Christians across the time divide share core convictions: God created everything, humans bear God's image, sin corrupted creation, Christ offers redemption. These essential truths don't depend on Earth's age. The Gospel remains powerful whether preached in a 6,000 or 6-billion-year-old universe.

Distinguishing core from peripheral doesn't minimize the time question but properly contextualizes it. Young-earth believers needn't doubt others' salvation over geological ages. Old-earth believers shouldn't dismiss young-earth faith as necessarily anti-intellectual. The Body of Christ includes members with different perspectives on peripheral issues while uniting around central truths.

This approach allows learning from each other. Young-earth creationists' commitment to biblical authority challenges compromise with secular assumptions. Their research into rapid geological processes and soft tissue preservation pushes scientific boundaries. Old-earth believers' engagement with mainstream science demonstrates faith's intellectual viability and reveals God's glory in deep time. Both perspectives contribute to understanding God's truth.

Practical Implications

For Science Education

How should Christians approach science education given chronological controversies? Several principles help:

First, understand what mainstream science teaches and why. Even if ultimately disagreeing, knowing conventional geology's logic enables informed critique. Ignorance serves neither truth nor faith. Students should learn radiometric dating's methods, assumptions, and applications.

Second, distinguish observations from interpretations. Rock layers exist—their age is interpreted. Isotope ratios can be measured—their implications are debated. Teaching this distinction develops critical thinking useful beyond origins debates.

Third, present multiple perspectives respectfully. Rather than indoctrinating with one view, explore different approaches' strengths and weaknesses. Let students wrestle with evidence and arguments. Faith grows stronger through honest engagement than sheltered isolation.

For Church Unity

Time disputes have divided churches and damaged witness. How can communities maintain unity amid disagreement?

Practice theological triage—distinguish essential from important from peripheral. Make space for different views on peripheral issues while maintaining standards on essentials. A church might include both young and old-earth believers while requiring affirmation that God created.

Foster respectful dialogue. Create forums for discussing controversial issues charitably. Model how Christians can disagree agreeably. Emphasize learning from those with different perspectives rather than converting them.

Focus on shared mission. Whatever Earth's age, people need the Gospel. Let evangelism and discipleship unite where chronology divides. Dinosaurs matter less than souls.

The Continuing Journey

The question of time remains unresolved and likely will until Christ returns. New discoveries—whether soft tissue in fossils or refined dating techniques—will provide fresh evidence requiring interpretation. Thoughtful Christians will continue wrestling with how to understand both God's word and God's world.

This ongoing tension needn't discourage but can inspire. Each generation engages these questions with new knowledge and perspectives. The conversation itself honors God by taking both revelation and creation seriously. Whether dinosaurs lived millions of years ago or thousands, they accomplished God's purposes in their time.

Our next chapter examines how dinosaurs' time ended—through asteroid impact, volcanic catastrophe, or divine judgment. The extinction event that eliminated dinosaurs raises its own theological questions about God's sovereignty over creation and destruction. However old Earth's rocks, they record a massive dying that demands explanation and reflection.

For Reflection

1. What evidence regarding Earth's age do you find most compelling? Most challenging to your position?
2. How do you balance confidence in your view with humility about what we don't know?
3. How can churches maintain unity when members disagree about the age of Earth?

Did You Know?

Radiocarbon dating of dinosaur bones has become a battleground between young and old-earth scientists. When creation researchers submit samples to dating labs without revealing their source, they often get results. When labs learn the samples are dinosaur bones, they sometimes refuse testing or dismiss results as contamination. This gatekeeping frustrates young-earth researchers who argue science should follow evidence wherever it leads.

Consider This

The oldest verified tree, a Great Basin bristlecone pine named Methuselah, is nearly 5,000 years old. Tree ring chronologies extended through overlapping patterns reach back about 12,000 years. These provide independent calibration for carbon-14 dating and challenge both extreme young-earth chronologies placing creation at 6,000 years and appearance-of-age arguments. Real trees record real years, requiring honest engagement with time's mysteries.

For Further Study

* Snelling, Andrew. Earth's Catastrophic Past
* Young, Davis and Stearley, Ralph. The Bible, Rocks and Time
* Ross, Hugh. A Matter of Days
* Wise, Kurt. Faith, Form and Time
* Mortenson, Terry and Ury, Thane, eds. Coming to Grips with Genesis

Chapter 10:

The Great Extinction: Asteroid or Flood?

"He looks on the earth and it trembles; he touches the mountains and they smoke!" — Psalm 104:32

The Day the Earth Burned

Luis Alvarez was a Nobel Prize-winning physicist, not a paleontologist. But in 1980, he and his geologist son Walter revolutionized our understanding of dinosaur extinction. While studying rock layers at the K-T boundary near Gubbio, Italy, they discovered something extraordinary: a thin clay layer enriched with iridium, an element rare on Earth but common in asteroids.

This iridium anomaly appeared worldwide at precisely the geological moment dinosaurs vanished. The Alvarezes proposed a radical hypothesis: a massive asteroid impact 66 million years ago ended the Mesozoic Era in global catastrophe. The scientific community initially resisted—gradual climate change or volcanic activity seemed more plausible than cosmic catastrophe. But evidence mounted relentlessly.

In 1990, researchers identified the smoking gun: Chicxulub crater, buried beneath Mexico's Yucatan Peninsula. This 110-mile-wide scar matched the predicted size, age, and location for the dinosaur killer. Computer models revealed the impact's horrific consequences: global wildfires, tsunamis 1,000 feet high, and an "impact winter" as debris blocked sunlight for months. The age of dinosaurs ended not with a whimper but with an apocalyptic bang.

Yet for many Christians, this secular catastrophe story rings hollow. Where is God in this random cosmic accident? How does this fit with biblical history? And could the greatest mass death in Earth's history instead point to the greatest judgment—Noah's flood?

The Asteroid Impact Evidence

Iridium and Shocked Quartz

The evidence for asteroid impact is compelling. The iridium layer appears at over 100 sites worldwide, always at the K-T (now K-Pg) boundary. The amount of iridium suggests an asteroid 6-10 miles in diameter. This cosmic bullet delivered energy equivalent to 100 million megatons of TNT—billions of times more powerful than all human nuclear weapons combined.

Beyond iridium, the boundary layer contains shocked quartz—crystals deformed by incredible pressure only found at impact sites and nuclear test grounds. Tiny glass spherules formed from rock melted and ejected by impact appear globally. The layer even contains chromium isotope ratios matching certain meteorites but not Earth rocks.

The Chicxulub crater itself tells a violent story. Drill cores reveal a sequence of catastrophic deposits: first, huge blocks of shattered bedrock from the impact. Then a thick layer of breccia—jumbled rock fragments—from the collapsing crater walls. Above this, sediments containing impact spherules and finally the iridium-rich clay. The entire sequence speaks of violence beyond imagination.

The Killing Mechanisms

How exactly did an asteroid kill dinosaurs globally? The impact alone, while devastating locally, couldn't directly harm creatures on the opposite side of Earth. Instead, cascading consequences created a perfect storm of destruction.

First came the thermal pulse. Impact ejecta reentering the atmosphere heated the entire sky to oven temperatures for hours. Computer models suggest surface temperatures briefly reached 500°F globally. Forests ignited spontaneously. Any unsheltered creature was literally cooked alive. Soot in the boundary layer confirms widespread wildfires.

Next, darkness fell. Vaporized rock and soot blocked sunlight for months, possibly years. Photosynthesis ceased. Plants died, followed by herbivores, then carnivores. Ocean food chains collapsed as phytoplankton perished. The impact winter resembled nuclear winter scenarios, but worse and longer-lasting.

Acid rain followed as impact-generated nitrogen oxides formed nitric acid in the atmosphere. Ocean surface waters acidified, dissolving shells of marine organisms. The geological record shows mass extinction of foraminifera and other shelled creatures precisely at the boundary.

Long-term climate chaos ensued. After the impact winter, greenhouse gases from vaporized limestone caused extreme warming. Wild temperature swings stressed surviving ecosystems for thousands of years. The world that eventually emerged looked nothing like the one destroyed.

Selectivity of Extinction

The extinction pattern provides crucial clues. Dinosaurs died out completely (except birds), along with pterosaurs, marine reptiles, and ammonites. Yet crocodiles, turtles, mammals, and birds survived. Why such selectivity?

Body size mattered—nothing over 55 pounds survived on land. Smaller animals could shelter more easily and required less food during the crisis. Aquatic creatures fared better, buffered by water from temperature extremes. Burrowing animals had refuge underground. Seeds and spores could remain dormant until conditions improved.

This selective survival matches impact consequences. Creatures dependent on continuous food supplies died when ecosystems collapsed. Those capable of dormancy, scavenging, or eating detritus survived. The pattern makes ecological sense given a sudden catastrophe followed by prolonged environmental stress.

The Flood Alternative

Catastrophism Revisited

Young-earth creationists see the same evidence but interpret it radically differently. The K-T boundary represents not an asteroid impact 66 million years ago but a particular stage of Noah's flood about 4,500 years ago. The global distribution of catastrophic deposits fits perfectly with a worldwide deluge.

Dr. Steve Austin, with a PhD in geology from Penn State, argues the iridium anomaly has multiple possible sources. Volcanic eruptions can concentrate iridium. The Deccan Traps in India—massive volcanic deposits roughly contemporaneous with dinosaur extinction—could provide iridium without requiring an asteroid. More intriguingly, creation scientists propose that the "fountains of the great deep" breaking up (Genesis 7:11) involved massive volcanism that could deposit iridium globally.

The selectivity of extinction also fits flood geology. Dinosaurs and marine reptiles, being large and air-breathing, would struggle in turbulent flood waters. Mammals and birds, being more mobile and intelligent, could temporarily escape rising waters, explaining their higher survival rate. The pattern reflects not random asteroid consequences but hydraulic sorting by mobility and habitat.

Flood geologists also note that dinosaur fossils often occur in massive bonebeds suggesting catastrophic burial. The Dinosaur Provincial Park in Alberta contains thousands of specimens jumbled together. The Morrison Formation across the American West shows similar patterns. These look like flood deposits, not gradual accumulation.

Rapid Burial Evidence

The exceptional preservation of many dinosaur fossils requires explanation. Articulated skeletons with bones in life position indicate rapid burial before scavengers and decay could scatter remains. Soft tissue preservation, including Mary Schweitzer's famous T. rex blood vessels, suggests minimal time between death and fossilization.

Consider fossilized dinosaur eggs, embryos, and nests. In China's Liaoning Province, Oviraptor fossils preserve adults brooding nests, apparently overwhelmed while protecting eggs. Such behavior fossils require virtually instantaneous burial. Normal decay would destroy such delicate arrangements in days.

Dinosaur trackways present another puzzle. Thousands of footprints preserve fine details like skin impressions, requiring rapid covering by sediment. Yet the same formations often contain multiple track layers. Flood geologists propose oscillating water levels during the flood—dinosaurs desperately seeking higher ground between tsunami-like waves, leaving tracks subsequently buried by next surge.

The Liscomb Bone Bed in Alaska contains thousands of Edmontosaurus bones from juveniles to adults, disarticulated but showing little weathering. The standard interpretation invokes seasonal mortality during migration. But flood geologists see a herd overwhelmed by catastrophic flooding, bodies transported short distances before rapid burial.

Problems with Impact Theory

Young-earth creationists identify several challenges to the asteroid hypothesis:

First, the Deccan Traps represent one of the largest volcanic events in Earth history, roughly concurrent with dinosaur extinction. These eruptions could produce many impact signatures—iridium, spherules, shocked minerals—without requiring an asteroid.

Why invoke an unnecessary extraterrestrial cause?

Second, dinosaur decline arguably began before the K-T boundary. Late Cretaceous diversity was decreasing, suggesting gradual extinction rather than sudden catastrophe. The asteroid becomes merely the final blow to already struggling populations. The flood model better explains progressive burial of different ecological zones.

Third, some dinosaur fossils appear above the K-T boundary, though mainstream scientists dismiss these as reworked specimens. In China and Montana, apparently fresh dinosaur bones occur in Paleocene deposits. If the impact ended dinosaurs instantly, such finds shouldn't exist. But if the boundary represents a flood horizon, post-boundary dinosaurs make sense—some survived initially before later local catastrophes.

Volcanic Catastrophe: A Third Option

The Deccan Traps

Whether accepting conventional dating or young-earth chronology, the Deccan Traps demand attention. These layered volcanic deposits in India cover 200,000 square miles and reach 6,000 feet thick in places. By any timeline, they represent catastrophic volcanism roughly contemporaneous with dinosaur extinction.

Conventional geologists date the eruptions over about 800,000 years, with the most intense phase coinciding with the K-T boundary. The eruptions released enormous amounts of sulfur dioxide and carbon dioxide, causing acid rain and climate chaos. Some researchers argue this volcanism alone could have caused mass extinction without asteroid assistance.

Young-earth geologists compress the timeframe but agree on the catastrophic nature. They see the Deccan Traps as part of the flood's tectonic violence—the "fountains of the great deep" breaking up. Such massive volcanism would indeed cause many features attributed to asteroid impact while fitting the biblical narrative of global catastrophe.

The volcano theory gains support from other mass extinctions apparently coinciding with large volcanic provinces. The Permian extinction aligns with the Siberian Traps. The Triassic extinction matches the Central Atlantic Magmatic Province. Perhaps Earth's internal violence, not cosmic accidents, drives mass extinctions.

Multiple Causes?

Increasingly, researchers recognize that monocausal explanations may be insufficient. Perhaps dinosaur extinction resulted from multiple stresses—volcanic winter followed by asteroid impact, or vice versa. Marine regression exposing continental shelves, climate change from continental drift, and ecological disruption from angiosperm evolution all preceded the final catastrophe.

From a theological perspective, multiple causes need not mean random coincidence. God's sovereignty encompasses all secondary causes. Whether through asteroid, volcano, or flood—or all three—God orchestrated events according to His purposes. The means matter less than the meaning.

This multicausal approach resonates with biblical descriptions of divine judgment employing multiple instruments. The plagues of Egypt involved natural phenomena—water turning red, frogs, locusts—orchestrated supernaturally. Babylon fell through military conquest serving divine purposes. God works through natural means without being limited by them.

Theological Reflections on Extinction

Divine Sovereignty Over Catastrophe

However dinosaurs died, their extinction raises profound theological questions. Does God cause mass extinctions? Allow them? Use them for greater purposes? The scale of death—not just dinosaurs but 75% of all species—challenges easy theodicies.

Scripture consistently affirms God's sovereignty over natural catastrophes. He sends floods (Genesis 6-9), fire from heaven (Genesis 19), and plagues (Exodus 7-12). Jesus calmed storms, demonstrating authority over chaotic nature. Yet Scripture also distinguishes God's active will from permissive will. Did God hurl the Chicxulub asteroid or merely allow cosmic processes He established to unfold?

The extinction's outcome suggests purpose beyond randomness. Dinosaur elimination opened ecological space for mammalian radiation, eventually enabling human existence. From our perspective, the catastrophe appears providential. But this anthropocentric interpretation risks minimizing the real suffering of billions of creatures.

Patterns of Death and Renewal

The fossil record reveals multiple mass extinctions throughout Earth history. The Permian extinction eliminated 90% of marine species. The Ordovician, Devonian, and Triassic events each reshaped life dramatically. After each dying, life resurged with new forms and greater complexity. Death enabled creativity.

This pattern resonates with biblical themes. The flood narrative explicitly links judgment with new creation. Israel's exile enabled spiritual renewal. Most profoundly, Christ's death brought resurrection life. Could mass extinctions reflect this deep pattern—death as prelude to greater life?

Yet we must avoid glibly spiritualizing real suffering. Each extinct species was unique, irreplaceable. Tyrannosaurus rex will never hunt again. Quetzalcoatlus will never soar. The particularity of loss matters. God remembers sparrows; surely He remembers Microraptor.

Eschatological Perspectives

Dinosaur extinction offers sobering perspective on Earth's fragility and our own mortality. If such magnificent creatures could vanish suddenly, what guarantees human permanence? The same processes that ended dinosaurs—impacts, volcanism, climate change—continue threatening.

Some Christians see dinosaur extinction as prefiguring final judgment. As the Mesozoic ended in fire, so will the present age (2 Peter 3:10). As a remnant survived to inherit renewed Earth, so will God's people inherit new creation. The pattern of judgment and renewal continues until ultimate fulfillment.

Others find hope in extinction's incompleteness. Birds survived as dinosaurian descendants, transformed for new worlds. Similarly, death cannot ultimately triumph. Creation will be transformed, not annihilated. Perhaps in new creation we'll discover what dinosaurs were always meant to become.

Living with Catastrophe

Scientific and Faith Perspectives

Christians engaging dinosaur extinction must navigate between scientific evidence and theological conviction. The asteroid impact has compelling evidence, yet seems to reduce history to cosmic accident. The flood explanation preserves divine purpose but requires rejecting mainstream geology. Volcanic theories offer middle ground but satisfy neither camp fully.

Perhaps the way forward involves holding multiple perspectives in creative tension. The physical mechanism—asteroid, volcano, or flood—matters less than recognizing God's sovereignty over all. Young-earth believers can acknowledge impact evidence while maintaining recent timescales. Old-earth believers can see divine providence in cosmic catastrophes.

This approach requires intellectual humility. We weren't there to observe dinosaur extinction. Our reconstructions from limited evidence remain provisional. New discoveries regularly revise theories. Dogmatism about extinction mechanisms serves neither science nor faith.

Practical Implications

How does dinosaur extinction inform Christian living? Several lessons emerge:

First, creation's contingency reminds us of proper priorities. Magnificent creatures that ruled Earth for 160 million years vanished suddenly. Human achievements and concerns shrink to proper proportions. We're temporary residents on a fragile planet.

Second, God's sovereignty encompasses catastrophe. Whether through natural processes or direct intervention, nothing escapes divine providence. This doesn't explain suffering but locates it within larger purposes we glimpse only partially.

Third, extinction and renewal reveal creation's dynamism. God didn't create static perfection but unfolding drama. Change, even catastrophic change, serves creative purposes. Our present trials may enable unimaginable futures.

Finally, studying extinction cultivates appropriate humility about human knowledge and control. Despite scientific advances, we remain vulnerable to forces beyond mastery. Wisdom lies in recognizing our limits while fulfilling our calling as creation's stewards.

The Story Continues

Dinosaur extinction marked not an ending but transformation. From disaster's ashes arose the age of mammals, eventually producing creatures capable of studying their predecessors. We exist because dinosaurs don't—a sobering reminder of life's interconnectedness and contingency.

Yet questions remain. Why did God permit or cause such massive dying? How does extinction serve divine purposes? What does this mean for Earth's future and humanity's role? These questions push beyond paleontology into philosophy and theology's depths.

Our next chapter examines one of the most intriguing aspects of dinosaur science—the discovery that some dinosaurs bore feathers and gave rise to birds. This evolutionary transition raises fascinating questions about classification, transformation, and the boundaries between created kinds. In exploring feathered dinosaurs, we engage directly with evidence for evolutionary change while seeking to understand it within a framework of faith.

For Reflection

1. Which extinction theory do you find most convincing? Why?
2. How does massive extinction fit with your understanding of God's character?
3. What lessons does dinosaur extinction offer for contemporary environmental concerns?

Did You Know?

The Chicxulub impact created a ring of cenotes (water-filled sinkholes) around the crater's edge in Mexico's Yucatan Peninsula. These beautiful swimming holes, sacred to the Maya, mark the buried rim of Earth's most catastrophic impact. From mass death came springs of life-giving water–an unintended parable of redemption from destruction.

Consider This

Some dinosaurs were fossilized in sleeping positions, like Mei long ("sleeping dragon") found in China with its head tucked under its wing like a modern bird. Whether they died peacefully and were gently buried or were overwhelmed while sleeping by catastrophic events, such fossils remind us that dinosaurs lived ordinary lives between the dramatic moments that paleontology typically preserves.

For Further Study

* Alvarez, Walter. T. rex and the Crater of Doom
* Courtillot, Vincent. Evolutionary Catastrophes
* Oard, Michael. Dinosaur Challenges and Mysteries
* Archibald, J. David. Dinosaur Extinction and the End of an Era
* Powell, James. Night Comes to the Cretaceous

Chapter 11:

Feathered Dinosaurs and the Origin of Birds

"Even the sparrow finds a home, and the swallow a nest for herself, where she may lay her young, at your altars, O Lord of hosts, my King and my God."
— Psalm 84:3

The Farmer's Revolutionary Discovery

Li Yinfang was splitting slate for his farmhouse floor in China's Liaoning Province when the rock revealed an impossible creature. The year was 1996, and between the stone layers lay a small dinosaur covered in what looked like fur or primitive feathers. Sinosauropteryx, as scientists would name it, shattered the neat divisions between dinosaurs and birds that had stood since Darwin's day.

Within months, the farming villages of Liaoning became ground zero for a paleontological gold rush. Farmers who once earned pennies per day could sell exceptional fossils for thousands of dollars. The hills yielded treasure after treasure: four-winged Microraptor, the sleeping dragon Mei long, and dozens of other feathered dinosaurs that would revolutionize our understanding of bird origins.

These discoveries arrived at a pivotal moment in the creation-evolution debate. For young-earth creationists, birds and dinosaurs were separate created kinds. For advocates of intelligent design, the complexity of feathers argued against gradual evolution. For theistic evolutionists, these fossils revealed God's creative process in action. The feathered dinosaurs of China would test each position's explanatory power.

The Archaeopteryx Enigma

The First Bird?

Before Liaoning's feathered dragons, Archaeopteryx stood alone as evolution's icon. Discovered in 1861 in Germany's Solnhofen limestone, this crow-sized creature possessed a mosaic of features: feathered wings like a bird, teeth and tail like a reptile. Darwin's supporters immediately claimed it as proof of evolutionary transition.

The timing seemed providential—or suspiciously convenient, depending on perspective. Just two years after Origin of Species, here was a perfect intermediate form. Thomas Henry Huxley, Darwin's bulldog, used Archaeopteryx to argue that birds descended from dinosaurs. Owen and other critics countered that it was simply a bird with some reptilian features, not evidence of transformation.

For over a century, Archaeopteryx remained contentious yet isolated. Where were the other transitional forms? If birds evolved from dinosaurs, shouldn't we find a gradual sequence of increasingly bird-like dinosaurs? The absence of such fossils became a creationist talking point—until Liaoning changed everything.

Design Perfection in Ancient Feathers

Archaeopteryx's feathers prove remarkably modern in structure. Under microscopic examination, they show asymmetric vanes essential for flight, hierarchical branching from rachis to barbs to barbules, and even melanosomes indicating black coloration like modern ravens. These aren't primitive proto-feathers but fully functional flight equipment.

From an intelligent design perspective, such complexity appearing suddenly argues against gradual evolution. Michael Benton calculated that a functional feather requires precise coordination of multiple proteins and developmental pathways. Random mutations would more likely produce useless fluff than aerodynamic surfaces. The mathematical improbability of evolving feathers by chance becomes astronomical.

Yet theistic evolutionists see divine creativity working through natural processes. God front-loaded evolution with potential for feathers, activated when environmental pressures favored them. The fact that some dinosaurs developed feathers for display or insulation before flight shows multiple pathways to the same innovation—evidence of robust design principles rather than random wandering.

The Chinese Revolution

Feathers Everywhere

Liaoning's fossil beds, formed from volcanic ash that buried creatures instantly, preserve extraordinary detail. The discoveries began with Sinosauropteryx's simple filaments and escalated rapidly. Protarchaeopteryx showed more complex feathers. Caudipteryx displayed full plumage. Microraptor possessed four wings with asymmetric flight feathers.

The diversity astonished researchers. Some dinosaurs bore simple down-like fuzz for insulation. Others showed symmetric decorative plumes. Still others had fully modern asymmetric flight feathers. The variety suggested feathers evolved multiple times for different functions—insulation, display, and eventually flight.

Even more surprising was the phylogenetic distribution. Feathers appeared not just in creatures close to birds but throughout theropod dinosaurs. Yutyrannus, a 30-foot tyrannosaur, bore feathers. Baby Tyrannosaurus might have been fuzzy. The question shifted from "Which dinosaurs had feathers?" to "Which didn't?"

Four-Winged Dragons

Microraptor deserves special attention as perhaps the most challenging fossil for traditional categories. This crow-sized predator possessed fully developed flight feathers on all four limbs. Its leg feathers weren't vestigial but functional, creating a biplane configuration unknown in modern animals.

How do we interpret such creatures? Young-earth creationists initially dismissed Chinese feathered dinosaurs as frauds, remembering the embarrassing "Archaeoraptor" hoax. But as specimens multiplied and underwent rigorous analysis, denial became impossible. These were real creatures that didn't fit neat categories.

Some creationists propose Microraptor as a separately created kind that happened to share features with both dinosaurs and birds. But this multiplication of kinds strains credulity. Others suggest these represent post-flood diversification within kinds—perhaps bird descendants that reevolved reptilian features. But the anatomical details argue against this scenario.

Intelligent design advocates see Microraptor as evidence of creative exploration rather than evolutionary transition. Why would evolution produce four-winged creatures if two wings work fine? The variety suggests a designer experimenting with different solutions, not natural selection climbing a single peak.

Colors from the Cretaceous

Breakthrough techniques analyzing melanosomes—pigment-bearing structures—in fossilized feathers revealed dinosaur colors. Sinosauropteryx showed reddish-brown stripes. Anchiornis displayed a black body with reddish head crest. Borealopelta, though unfeathered, preserved intricate camouflage patterns.

These discoveries humanize dinosaurs remarkably. They weren't just bones but living creatures with colors serving specific functions—camouflage from predators, display for mates, species recognition. The level of preservation allowing such analysis seems almost miraculous, whether viewed as divine providence or extraordinary fortune.

For young-earth creationists, such exquisite preservation suggests recent burial. How could delicate pigment structures survive millions of years? But conventional scientists note that melanosomes are remarkably stable, surviving in fossils over 100 million years old. The debate continues over preservation mechanisms and their implications for age.

Birds: Living Dinosaurs?

The Classification Crisis

Modern phylogenetic analysis places birds firmly within dinosaurs, specifically as maniraptorans closely related to Velociraptor and Deinonychus. The anatomical evidence is overwhelming: hollow bones, three-toed feet, wishbones, and numerous other shared features. Genetically, birds are dinosaurs as surely as bats are mammals.

This reclassification bothers many Christians. Genesis describes God creating birds on Day Five and land animals on Day Six. If birds are dinosaurs, did God create them twice? The taxonomic revision seems to challenge biblical categories and chronology.

Various responses emerge. Some maintain that biblical "kinds" don't correspond to modern taxonomy—the Hebrew terms describe functional groups, not biological clades. Others propose that Genesis describes God's creative acts from an earth-surface perspective, not strict chronology. Still others accept the tension without resolution, trusting that apparent conflicts stem from incomplete understanding.

Living Fossils

Birds retain numerous dinosaurian features that seem inexplicable except through ancestry. They walk on their toes like theropods. Their ankles bend like dinosaur ankles, not like mammals or lizards. The embryonic hand shows three digits matching theropod anatomy before fusing into the wing.

Most remarkably, chickens retain genetic potential for dinosaurian features. Scientists have coaxed chicken embryos to develop teeth, long tails, and unfused hand digits—all dinosaur characteristics. This "evo-devo" research suggests birds carry their evolutionary history in suppressed genes, activatable under proper conditions.

How do different perspectives interpret this evidence? Young-earth creationists see common design rather than common descent—God used similar genetic toolkits for different creatures. ID proponents emphasize that reactivating existing genes differs vastly from evolving new features. Theistic evolutionists see God's economy, preserving ancient potentials for possible future use.

The Origin of Flight

From the Ground Up or Trees Down?

How did dinosaurs take to the air? Two main hypotheses compete. The "ground up" model envisions small theropods running and jumping, with proto-wings providing thrust or stability. The "trees down" model imagines climbers gliding between branches, gradually developing powered flight.

Evidence supports both scenarios. Some feathered dinosaurs like Caudipteryx were clearly terrestrial, using wings for display or brooding. Others like Microraptor show adaptations for climbing and gliding. Perhaps different lineages achieved flight through different routes—another example of convergent evolution or divine creativity exploring multiple solutions.

The biomechanics of flight origin fascinate engineers. Microraptor's four-winged configuration actually works—models demonstrate stable gliding and possibly powered flight. Yi qi, a bat-winged dinosaur with membrane rather than feathers, represents yet another flight experiment. The diversity suggests flight's attractions led to multiple attempts with varying success.

Irreducible Complexity Revisited

Flight seems irreducibly complex—requiring coordinated changes in skeleton, muscles, nervous system, and integument. How could gradual evolution produce systems that only function when complete? Half a wing appears worse than useless, hindering running without enabling flight.

Yet Chinese fossils show apparent intermediates. Feathers first appeared for insulation or display, only later adapted for flight. Wing-assisted incline running—using proto-wings for traction climbing slopes—provides immediate benefit before flight. Each stage offered advantages, making the next stage achievable.

ID advocates respond that functional intermediates don't explain origin of information required for transitions. Where did genetic instructions for feather development originate? How did regulatory networks reorganize for flight? Finding functional intermediates addresses only part of the challenge. The deeper mystery involves origin of biological information itself.

Theological Implications

Created Kinds and Fluid Boundaries

Feathered dinosaurs challenge concepts of fixed "kinds." If birds descended from dinosaurs, where exactly is the boundary? Archaeopteryx was once clearly a bird; now many classify it as a dinosaur. Velociraptor was clearly a dinosaur; now we recognize it as close to bird ancestry. Categories blur uncomfortably.

Some young-earth creationists propose broader kinds than previously imagined. Perhaps the bird kind includes some creatures we call dinosaurs. But this stretches the kind concept toward meaninglessness. If Tyrannosaurus and hummingbirds belong to one kind, what doesn't?

Others maintain that similarities reflect common design, not descent. But why would God design dinosaurs with suppressed genetic potential for bird features? The pattern makes more sense given actual ancestry than independent creation with unused bird genes.

Beauty Beyond Survival

Many feathered dinosaurs display features seemingly excessive for survival. Epidexipteryx's tail ribbons, Similicaudipteryx's plume arrangements, and various crests and displays suggest aesthetics beyond utility. Why did dinosaurs develop such elaborate ornamentation?

Theistic evolutionists see divine creativity delighting in beauty for its own sake. Evolution becomes God's palette for artistic expression, not mere survival optimization. The excess reveals personality behind process—a God who values splendor, not just efficiency.

This perspective enriches our view of creation's purpose. Dinosaurs didn't exist merely as evolutionary stepping stones or ecological functionaries. They lived as creatures valuable in themselves, displaying divine artistry. Their feathers declare God's glory as surely as stars and mountains.

Transformation and Hope

The dinosaur-bird transition offers powerful metaphors for Christian life. What seemed like ending—dinosaur extinction—became new beginning as birds filled empty skies. Death and resurrection patterns permeate creation's history.

Paul speaks of creation groaning for redemption, subject to futility yet hoping for transformation (Romans 8). Feathered dinosaurs embody this groaning—creatures caught between earth and sky, running and gliding, dying out yet living on transformed. They remind us that present struggles birth future glory.

The transformation wasn't instant but involved millions of years of intermediate forms. Similarly, sanctification involves lengthy process, often unclear in direction. We're intermediate forms ourselves—no longer merely human yet not fully glorified, bearing both earthly and heavenly natures.

Integration and Mystery

Multiple Perspectives, Shared Wonder

Christians examining feathered dinosaurs reach different conclusions. Young-earth creationists see separately created kinds with superficial similarities. Old-earth creationists see progressive creation with transitions. Theistic evolutionists see gradual transformation guided by providence. ID theorists see design however implemented.

Yet all can share wonder at these remarkable creatures. Microraptor's iridescent plumage, Sinosauropteryx's striped tail, Yi qi's bat-like wings—these inspire awe regardless of origin theories. The diversity, beauty, and strangeness of feathered dinosaurs expand our vision of divine creativity.

Perhaps God allows interpretive ambiguity to prevent intellectual pride. If evidence clearly supported only one view, we might trust our reasoning rather than God's revelation. Uncertainty keeps us humble, dependent on faith beyond empirical proof.

Questions for Further Research

Feathered dinosaurs raise questions science continues investigating:

How many times did feathers evolve independently? Current evidence suggests possible multiple origins, challenging simple linear evolution. Did flight evolve once or several times? Each new discovery reveals greater complexity.

What killed non-avian feathered dinosaurs while birds survived? Size alone doesn't explain the pattern—some extinct species were smaller than surviving birds. Did behavioral differences, metabolic adaptations, or providence determine survival?

Will we find feathered dinosaurs outside China? Exceptional preservation requires specific conditions, but absence of evidence isn't evidence of absence. Future discoveries may reveal global distribution of feathered dinosaurs.

Looking Forward

Feathered dinosaurs exemplify how scientific discoveries challenge and enrich faith. They force examination of assumptions about created kinds, evolutionary mechanisms, and biblical interpretation. Rather than threatening faith, they invite deeper engagement with both Scripture and nature.

Our next chapter shifts from scientific controversies to personal testimonies. How do Christian paleontologists integrate faith with fossil study? What wisdom have they gained from careers spent examining God's ancient works? Their stories provide practical guidance for anyone seeking to honor God while pursuing scientific truth.

The journey from terrible lizards to flying birds reminds us that God's ways transcend human categories. He works through processes we're only beginning to understand, creating wonders beyond imagination. In studying feathered dinosaurs, we glimpse divine creativity that transforms scales into feathers, death into flight, extinction into endless song.

For Reflection

1. How do feathered dinosaurs challenge or confirm your understanding of creation?
2. What does the dinosaur-bird transition suggest about God's creative methods?
3. How can Christians disagree about origins while maintaining unity in faith?

Did You Know?

The protein keratin forms both scales and feathers, requiring only regulatory changes to produce different structures. A simple mutation in chickens produces feathered feet instead of scaled, demonstrating how easily integument can transform. This biochemical relationship between scales and feathers suggests deep connections that transcend surface appearances.

Consider This

Confuciusornis, an early bird from China, shows remarkable sexual dimorphism–males possessed long tail ribbons absent in females. Thousands of specimens preserve evidence of ancient mating displays. These birds died at lake edges, possibly during breeding gatherings, creating fossil deposits recording behavior 120 million years old. Even in deep time, creatures sought beauty and companionship.

For Further Study

* Xu, Xing. Various papers on Chinese feathered dinosaurs
* Prum, Richard. The Evolution of Beauty
* Feduccia, Alan. Riddle of the Feathered Dragons (skeptical view)
* Dial, Kenneth. Papers on wing-assisted incline running
* Brusatte, Steve. The Rise and Reign of the Mammals

Chapter 12:

Christians Studying Dinosaurs

"Great are the works of the Lord, studied by all who delight in them." — Psalm 111:2

The Bone Hunter's Prayer

Dr. Mary Schweitzer knelt in the badlands of Montana, her hands covered in plaster and ancient rock dust. The Tyrannosaurus rex femur she was jacketing had broken during excavation, revealing its interior. What she saw made her gasp—the bone's center looked fresh, with flexible vessels and what appeared to be red blood cells. As a Christian paleontologist, she faced a moment that would define her career and test her faith.

"Lord," she prayed silently, "help me follow the evidence wherever it leads. Give me wisdom to understand what You're showing me through Your creation." That prayer began a journey that would revolutionize paleontology, challenge conventional wisdom about fossil preservation, and demonstrate how Christians can pursue cutting-edge science while maintaining vibrant faith.

Schweitzer's story exemplifies the experiences of Christian paleontologists worldwide—women and men who excavate dinosaurs, analyze fossils, and reconstruct ancient worlds while worshipping the God who created it all. Their testimonies provide unique insights into integrating faith with scientific rigor, handling controversy with grace, and finding God's glory in Earth's deep history.

Kurt Wise: The Price of Conviction

Choosing Scripture Over Science

Kurt Wise possessed everything needed for paleontological stardom. A Harvard PhD under Stephen Jay Gould, brilliant research on fossil floating forests, and job offers from prestigious universities. Yet he chose a different path, accepting a position at Bryan College—William Jennings Bryan's small Tennessee school—to develop creation science.

Wise's story begins with childhood fascination with fossils and firm biblical faith. As a teenager, he took scissors to his Bible, cutting out every verse that would have to go if evolution were true. When he finished, his Bible wouldn't hold together. That visual demonstration convinced him: he must choose between evolutionary science and biblical faith. He chose faith.

At Harvard, Wise never hid his beliefs. Gould, though an atheist, respected Wise's intelligence and integrity. Their relationship demonstrated that scientific disagreement needn't prevent personal respect. Wise produced excellent research on Paleozoic plants while maintaining young-earth convictions. His dissertation advisor called him one of his best students ever.

Creation Research

After Harvard, Wise dedicated his career to developing positive creation models rather than merely critiquing evolution. His work on floating forest ecosystems explains Carboniferous coal deposits without requiring millions of years. He proposes that pre-flood floating forests, destroyed and buried during the deluge, formed massive coal beds rapidly.

Wise's "discontinuity systematics" attempts to identify the original created kinds. By analyzing morphological gaps between organisms, he maps potential boundaries of biblical kinds. While controversial, his approach demonstrates sophisticated engagement with biological data from a creationist perspective.

His honesty about creation science's challenges earns respect even from opponents. Wise acknowledges that young-earth creationism currently lacks explanatory power compared to conventional geology. But he believes further research will vindicate biblical chronology. Meanwhile, he trains the next generation of creation scientists at Truett McConnell University.

Personal Cost and Conviction

Wise's choices came with significant personal cost. Former colleagues expressed disappointment that he "wasted" his Harvard training. He forfeited opportunities for mainstream research positions and funding. Some evangelicals criticize him for being too accommodating to conventional science. He stands between worlds, fully accepted by neither.

Yet Wise radiates joy in his calling. He sees his work as worship, exploring God's creation to understand divine wisdom. Every fossil testifies to God's creativity and judgment. His students catch his enthusiasm, learning to see dinosaurs as monuments to their Maker rather than evolutionary icons.

"If all the evidence in the universe turns against creationism," Wise once wrote, "I would be the first to admit it, but I would still be a creationist because that is what the Word of God seems to indicate." This statement horrifies some and inspires others, but it demonstrates remarkable intellectual honesty about the role faith plays in his scientific interpretations.

Mary Schweitzer: Soft Tissue and Hard Questions

Revolutionary Discoveries

Mary Schweitzer's journey to paleontology began unusually. A molecular biologist studying marine algae, she returned to college after homeschooling her children. Jack Horner's dinosaur class captivated her. When she mentioned her background, Horner immediately recruited her for molecular paleontology—a field that barely existed.

Schweitzer brought unique perspectives to fossil analysis. While traditional paleontologists studied morphology, she examined molecular structures. Her breakthrough came in 2005 when she discovered soft tissue in a T. rex femur. The paleontological community reacted with skepticism bordering on hostility. Soft tissue couldn't survive 68 million years—therefore, she must be wrong.

But Schweitzer persisted, refining techniques and expanding discoveries. She found similar preservation in multiple specimens—blood vessels, bone cells, and proteins. Each discovery pushed the boundaries of what scientists thought possible. Her work forced revision of fossilization theories and opened new avenues for studying extinct life.

Faith Under Fire

As an evangelical Christian, Schweitzer faced attacks from multiple directions. Young-earth creationists claimed her discoveries proved recent creation—if soft tissue couldn't last millions of years, dinosaurs must be young. Atheistic scientists suspected her faith biased her research toward impossible claims. She felt squeezed between camps.

Schweitzer maintains that science must follow evidence regardless of implications. The soft tissue is real; the radiometric dates are also real. Rather than choosing sides, she investigates preservation mechanisms that might allow organic molecules to survive deep time. Her proposed iron-facilitated preservation mechanism shows how blood chemistry might protect tissues far longer than expected.

Her approach models intellectual honesty. She doesn't claim her discoveries prove young earth or design, though some creationists wish she would. Neither does she deny faith's influence on her wonder at creation's complexity. She pursues truth wherever it leads, trusting that all truth is God's truth.

Integration in Practice

Schweitzer's laboratory practices demonstrate faith-science integration. She prays for wisdom before experiments. She thanks God for the privilege of studying His works. She maintains rigorous scientific standards while seeing spiritual significance in her research. Her students learn excellent technique alongside reverent wonder.

Her favorite Bible verse is Proverbs 25:2: "It is the glory of God to conceal a matter; to search out a matter is the glory of kings." Paleontology becomes royal work, uncovering secrets God embedded in creation. Each discovery reveals more of divine creativity and wisdom.

Schweitzer's advice to young Christians in science: "Don't check your brain at the church door or your faith at the laboratory door. God gave you both for good reasons. Use them together." This integration requires constant balance but yields rich rewards—deeper faith through scientific study and better science through faith-inspired wonder.

Arthur Chadwick: Adventist Paleontologist

Dinosaurs and Sabbath

Arthur Chadwick represents another stream of Christian paleontology—Seventh-day Adventist scientists who see Sabbath-keeping and creation faith as inseparable. Professor at Southwestern Adventist University, Chadwick combines rigorous field research with commitment to recent creation.

His work at the Hanson Ranch bone bed in Wyoming exemplifies careful methodology. Using GPS technology to map thousands of dinosaur bones, Chadwick documents one of North America's largest dinosaur graveyards. The jumbled remains of Edmontosaurus, Triceratops, and other species suggest catastrophic burial—consistent with flood geology.

Chadwick doesn't just collect data supporting his view. He engages mainstream paleontology seriously, publishing in peer-reviewed journals and presenting at professional conferences. His mapping techniques have been adopted by other researchers. He demonstrates that young-earth scientists can contribute valuable research while maintaining their convictions.

Educational Mission

Training the next generation motivates Chadwick's research. His field schools bring students to Wyoming badlands, teaching excavation techniques while discussing worldview implications. Students learn to identify fossils, jacket specimens, and interpret geological formations from a biblical perspective.

Chadwick emphasizes observational skills over ideological conformity. Students must accurately describe what they find before interpreting significance. This approach produces graduates who can engage mainstream science intelligently while maintaining creation faith. Some continue to graduate schools, carrying Adventist perspectives into wider academia.

His balanced approach acknowledges challenges facing flood geology while exploring solutions. How did ecosystems recover after the flood? Why do we find ordered fossil sequences? Rather than dismissing difficulties, Chadwick investigates them as research opportunities. This honesty strengthens rather than weakens student faith.

Peter Dodson: Catholic Perspectives

Sacramental Science

Peter Dodson brings Catholic sensibilities to dinosaur studies. Professor at University of Pennsylvania's veterinary school and world expert on ceratopsian dinosaurs, Dodson sees no conflict between evolutionary science and orthodox faith. His approach reflects Catholic tradition's comfort with natural theology.

Dodson's research spans continents, describing new species and analyzing dinosaur growth patterns. His work on Avaceratops revealed how baby dinosaurs' proportions changed dramatically during development. Such discoveries illuminate God's creative processes working through natural laws over deep time.

As a practicing Catholic, Dodson finds spiritual meaning in scientific work. Each fossil represents part of creation's grand narrative leading to Christ. Evolution becomes God's chosen method for creating diversity and beauty. The long history of life prepares for salvation history's culmination in the Incarnation.

Bridge Building

Dodson works to bridge divides between religious communities and mainstream science. He speaks at churches, explaining how Christians can accept evolution without abandoning faith. He challenges scientists to respect religious convictions rather than dismissing them as ignorance. His irenic approach opens dialogue where others create division.

His writing addresses pastoral concerns about evolution. How can Christians worship a God who creates through processes involving death and extinction? Dodson points to Christ's own suffering as God entering creation's pain. The Cross redeems not just human sin but all creation's groaning.

Dodson's integration appears effortless because it reflects centuries of Catholic thought harmonizing faith and reason. He stands on foundations laid by Augustine and Aquinas, who insisted truth cannot contradict truth. This theological tradition provides resources for engaging modern science constructively.

Common Themes and Divergent Paths

Shared Convictions

Despite theological differences, Christian paleontologists share core convictions:

Creation's Testimony: All see dinosaurs declaring God's glory. Whether created recently or evolving over millions of years, these magnificent creatures reflect divine creativity. Studying them becomes an act of worship.

Intellectual Integrity: Each strives for honesty in handling evidence. They may interpret data differently, but all commit to accurate observation and rigorous analysis. Faith doesn't excuse sloppy science.

Wonder and Humility: Dinosaurs inspire awe at God's works beyond human comprehension. This wonder breeds humility—recognition that our understanding remains partial, our theories provisional.

Vocational Calling: These scientists view their work as divine calling, not mere career. God equipped and called them to study His ancient works. This sense of vocation sustains them through challenges.

Different Approaches

Yet significant differences divide Christian paleontologists:

Age of Earth: Young-earth advocates like Wise and Chadwick maintain recent creation despite evidence for deep time. Old-earth scientists like Dodson accept conventional chronology. Schweitzer focuses on mechanisms rather than timescales.

Evolution's Role: Some reject macroevolution entirely, others accept it as God's creative method. This fundamental divide shapes how they interpret fossil evidence and understand life's history.

Biblical Interpretation: Approaches to Genesis vary from literal historical narrative to theological framework. These hermeneutical differences drive scientific disagreements about dinosaurs' place in Earth history.

Professional Engagement: Some work within mainstream science, others in parallel creationist institutions. This affects research opportunities, funding sources, and peer communities.

Wisdom for Aspiring Scientists

Practical Advice

Christian paleontologists offer guidance for students considering scientific careers:

Develop Excellence: Whatever your convictions about origins, pursue scientific excellence. Learn techniques thoroughly, understand mainstream theories completely, conduct research rigorously. Credibility requires competence.

Choose Mentors Wisely: Find advisors who respect your faith even if not sharing it. Gould mentored Wise despite their disagreements. Good mentors value integrity over ideological conformity.

Pick Battles Carefully: Not every hill deserves dying on. Focus on positive research contributions rather than constant confrontation. Build credibility through quality work before challenging paradigms.

Maintain Community: Connect with other Christians in science for support and accountability. Isolation breeds either compromise or extremism. Fellowship provides balance and encouragement.

Spiritual Disciplines

Integrating faith and fossil study requires intentional spiritual practices:

Regular Prayer: Begin fieldwork and laboratory sessions with prayer. Thank God for the privilege of studying His works. Ask wisdom for interpreting discoveries accurately.

Scripture Study: Ground your worldview in biblical truth while remaining open to creation's testimony. Let Scripture shape your approach to science without predetermining every conclusion.

Worship Through Work: See each fossil as cause for praise. Let wonder at dinosaurs' design lead to worship of their Designer. Transform scientific practice into spiritual discipline.

Humble Learning: Remain teachable from both Scripture and nature. Avoid dogmatism about uncertain matters. Hold convictions firmly but humbly, ready to refine understanding.

The Future of Christian Paleontology

Emerging Opportunities

New technologies create unprecedented opportunities for Christian paleontologists. CT scanning reveals internal structures without destructive analysis. Molecular techniques recover ancient proteins and possibly DNA. Computer modeling tests biomechanical hypotheses. Each advance opens fresh avenues for understanding God's ancient creatures.

Young Christians entering paleontology today inherit rich traditions from pioneers like Wise, Schweitzer, Chadwick, and Dodson. They need not choose between faith and science but can integrate both in service to truth. The field needs researchers who combine technical excellence with theological wisdom.

Continuing Challenges

Challenges remain for Christians in paleontology. Funding often depends on conformity to evolutionary paradigms. Publication requires navigating peer review potentially hostile to religious perspectives. Career advancement may conflict with convictional integrity.

Yet opportunities for faithful witness abound. As science confronts its limits—explaining consciousness, life's origin, fine-tuning—Christians offer resources from theological traditions. Dinosaur studies raise ultimate questions that purely naturalistic science cannot answer. Christian paleontologists stand ready to engage these deeper issues.

Conclusion: Faithful Witnesses

Christian paleontologists demonstrate that faith and fossil study can coexist fruitfully. Their diverse approaches show there's no single "Christian way" to study dinosaurs. What unites them is commitment to honor God through excellent science, wherever that leads.

Their testimonies encourage young Christians considering scientific careers. You need not abandon faith to study God's creation professionally. Whether maintaining young-earth convictions like Wise, pushing scientific boundaries like Schweitzer, or integrating evolution and faith like Dodson, you can pursue paleontology while following Christ.

These scientists model engagement with contested issues—not through culture war rhetoric but careful research, respectful dialogue, and humble faith. They show that Christians belong in paleontology's future, bringing unique perspectives to unlock creation's ancient mysteries.

Our next chapter turns from professional scientists to practical wisdom for everyday believers. When coworkers, children, or skeptics ask about dinosaurs, how should Christians respond? What answers honor both biblical faith and scientific evidence? The testimonies of Christian paleontologists provide foundation for faithful, thoughtful responses to dinosaur questions.

For Reflection

1. Which Christian paleontologist's approach most resonates with your own convictions? Why?
2. How can Christians in scientific fields maintain both professional excellence and spiritual integrity?
3. What can the broader Church learn from these scientists about engaging controversial issues?

Did You Know?

Jack Horner, who discovered Maiasaura and revolutionized understanding of dinosaur parenting, is dyslexic and never completed a college degree. Yet he became curator of paleontology at the Museum of the Rockies and inspired a generation of researchers. God uses unlikely people–including those with learning differences–to unlock creation's mysteries. Formal credentials matter less than passion and persistence.

Consider This

The Society of Vertebrate Paleontology has about 2,300 members worldwide. If even 10% are practicing Christians (a conservative estimate), over 200 believers professionally study dinosaurs and other fossils. These scientists quietly integrate faith and fossils daily, demonstrating that Christians contribute throughout paleontology despite stereotypes of universal secular materialism.

For Further Study

* Numbers, Ronald. The Creationists (historical context)
* Schweitzer, Mary. Various interviews and papers
* Wise, Kurt. Faith, Form, and Time
* Dodson, Peter. Interview in Dinosaurs: A Celebration
* American Scientific Affiliation: network of Christians in science

Chapter 13:

What to Say When Asked About Dinosaurs

"But in your hearts revere Christ as Lord. Always be prepared to give an answer to everyone who asks you to give the reason for the hope that you have. But do this with gentleness and respect." — 1 Peter 3:15

The Question at the Museum

Sarah watched her eight-year-old son Michael press his nose against the display case, mesmerized by the Allosaurus skeleton towering above them. The Natural History Museum buzzed with excited children on this Saturday afternoon, but Michael's question cut through the noise with laser precision: "Mom, our Sunday School teacher says God made everything in six days, but the sign says this dinosaur lived 150 million years ago. Who's lying?"

Sarah's stomach tightened. She'd dreaded this moment while simultaneously knowing it was inevitable. How could she honor both her son's faith and his love of science? How could she explain complexities that theologians and scientists debate without confusing or disillusioning her child? Her answer in this moment could shape Michael's relationship with both faith and learning for years to come.

This scene replays countless times in Christian families worldwide. Whether at museums, watching documentaries, or doing homework, dinosaur questions arise with uncomfortable regularity. These conversations matter deeply because they shape how the next generation understands the relationship between faith and science, Scripture and evidence, divine revelation and human discovery.

Understanding the Real Question

Listen for the Heart

When someone asks about dinosaurs and the Bible, they're rarely seeking technical discussions about radiometric dating or Hebrew grammar. The real questions usually run deeper:

"Can I trust the Bible?" lies beneath queries about Genesis and geology. The person wonders whether Scripture remains reliable when it seems to conflict with science. They're testing whether faith can withstand intellectual scrutiny.

"Is Christianity anti-science?" motivates many dinosaur discussions. People worry that following Christ means abandoning rational thought. They've heard stereotypes about Christians rejecting evidence, and they're checking whether faith requires intellectual suicide.

"How can I maintain faith in a scientific world?" drives questions from believers struggling to integrate Sunday sermons with Monday lectures. They love Jesus but also love learning. Dinosaurs become the test case for whether these loves can coexist.

Age-Appropriate Responses

Michael's question at age eight differs vastly from the same question at eighteen. Young children need simple, confidence-building answers. Teenagers require more nuance and respect for their developing critical thinking. Adults deserve comprehensive engagement with complexities.

For young children: "Different Christians understand dinosaurs differently. Some think God made them along with Adam and Eve. Others think God made them long before people. But all Christians agree that God made these amazing creatures! Isn't it wonderful that God is creative enough to make such incredible animals?"

For teenagers: "This is actually a great question that Christians have different views on. Let me share the main perspectives and why sincere believers hold them. What's most important is that God created everything, even if we're still figuring out exactly how and when."

For adults: "That touches on some fascinating discussions about how we interpret both Scripture and scientific evidence. There are several thoughtful Christian approaches to this question. Would you like to explore them together?"

Honest Answers to Tough Questions

"Doesn't Science Prove the Bible Wrong?"

This common question assumes science and Scripture necessarily conflict. A helpful response acknowledges the apparent tension while refusing the premise:

"I understand why it might seem that way. But many great scientists have been Christians who saw no conflict. The key is understanding what kind of book the Bible is. It's not a science textbook but God's revelation about who He is, who we are, and how we can know Him. Genesis teaches that God created everything good, humans are special, and sin brought death. These truths don't depend on whether creation took six days or six billion years."

Continue by explaining that Christians interpret Genesis differently:

- Some read it as literal history (young-earth view)

- Others see it as teaching theology through ancient literary forms (framework view)
- Still others believe God revealed truth in ways ancient people could understand (accommodation view)

Emphasize that sincere Christians disagree about how and when God created while uniting around the fact that He created.

"Were Dinosaurs on Noah's Ark?"

Children especially love this question, imagining Brachiosaurus barely squeezing through the ark's door. The answer depends on one's broader framework:

For those accepting young-earth chronology: "If dinosaurs lived alongside people, then yes, Noah would have taken them on the ark. But remember, he probably took young animals who were smaller and needed less food. Also, the Bible talks about 'kinds' of animals, which might be broader than our modern species. So Noah didn't need two of every dinosaur species, just representatives of each major kind."

For those accepting old-earth chronology: "Most Christians who accept geological timescales believe dinosaurs went extinct millions of years before humans appeared, so they wouldn't have been on the ark. Noah's flood might have been a local judgment on human sin rather than a global extinction event."

For those uncertain: "Christians have different views on this. What we know for sure is that God preserved life through Noah's obedience. Whether that included dinosaurs depends on when they lived, which Christians disagree about."

"Why Didn't God Mention Dinosaurs in the Bible?"

This question assumes the Bible should be encyclopedic, mentioning everything that exists. A thoughtful response reframes expectations:

"The Bible doesn't mention lots of things—penguins, kangaroos, bacteria, DNA. That's because the Bible has a specific purpose: revealing God's character and His plan for salvation. It tells us what we need to know about God and ourselves, not everything about the natural world."

You might add: "Some Christians think the Bible does mention dinosaurs. When Job talks about 'Behemoth' with a tail like a cedar tree, some see a description of a sauropod dinosaur. Others think it's poetic language for a hippopotamus or elephant. The Bible uses the language and concepts familiar to its original audience."

"How Can a Good God Create Killer Dinosaurs?"

Tyrannosaurus rex raises the problem of natural evil in particularly sharp teeth. This profound question deserves careful treatment:

"That's one of the deepest questions people ask—why is there suffering and death in God's good creation? Christians have wrestled with this for centuries. Some believe all death came after human sin, so originally even T. rex was vegetarian. Others think animal death was part of creation before sin, serving purposes we don't fully understand—like how predators today keep ecosystems balanced.

What we do know is that God isn't distant from suffering. In Jesus, God entered our world of pain and death. He experienced violence and overcame it through resurrection. Somehow, God is working even through difficult things to bring about His good purposes. The full answer awaits the new creation when God will make all things right."

Avoiding Common Pitfalls

Don't Overstate Your Case

Whether defending young-earth or old-earth positions, avoid claiming more certainty than evidence warrants. Overstatement damages credibility and faith:

Wrong: "Science has proven the earth is only 6,000 years old." Better: "Some scientists interpret evidence to support a young earth, though this remains a minority position."

Wrong: "Real Christians must accept evolution." Better: "Many faithful Christians accept evolution as God's method of creation, while others disagree."

Intellectual honesty builds trust. Acknowledge strengths in other positions and weaknesses in your own. This models humility and shows that faith doesn't require absolute certainty about everything.

Don't Create False Dilemmas

Avoid forcing unnecessary choices between faith and science:

Harmful: "You can either believe the Bible or believe in dinosaurs." Helpful: "All Christians believe God created dinosaurs. We disagree about when and how."

Harmful: "Evolution is atheistic religion masquerading as science." Helpful: "While some use evolution to argue against God, many Christians see it as God's method of creating."

False dilemmas push people away from faith by making it seem incompatible with obvious realities. Instead, show how faith encompasses and transcends scientific questions.

Don't Mock or Dismiss

Respect questioners and their concerns:

Dismissive: "Why worry about dinosaurs? Just focus on Jesus." Respectful: "I can see this really matters to you. Let's think through it together."

Mocking: "Only brainwashed people believe in millions of years." Respectful: "I understand why the evidence seems to point that way, though I interpret it differently."

People asking about dinosaurs often struggle with deeper faith questions. Dismissing their concerns suggests Christianity can't handle scrutiny. Patient engagement demonstrates confidence in truth.

Positive Approaches

Focus on the Gospel

Keep conversations centered on what matters most:

"These are fascinating questions, and Christians have different views about dinosaurs. But what unites us is far more important—God loves us so much that He sent Jesus to restore our broken relationship with Him. Whether dinosaurs lived millions of years ago or thousands, Christ died and rose for our salvation."

This doesn't dismiss the question but properly contextualizes it. Origins matter, but not as much as our eternal destiny. Help people see that they can follow Christ while continuing to explore questions about creation.

Celebrate Creation's Wonders

Use dinosaur discussions to inspire worship:

"Isn't it amazing that God created such incredible creatures? Whether you think Argentinosaurus lived recently or long ago, imagine the creativity required to design a 100-ton animal! The same God who crafted Microraptor's four wings crafted you with infinite care. Dinosaurs show us a God of boundless imagination and power."

This approach turns potential conflict into occasion for praise. Let dinosaurs do what creation does best—declare God's glory.

Encourage Further Study

Don't pretend to have all answers:

"That's a great question that deserves more time than we have right now. Here are some resources representing different Christian perspectives. I'd love to hear what you discover and discuss it further."

Providing balanced resources shows confidence that truth can withstand investigation. It also respects the questioner's intelligence and initiative.

Special Situations

At Museums and Schools

When Christian families visit natural history museums or children learn about dinosaurs in school, preparation helps:

Before visiting: "We're going to see amazing things God created. The museum will tell us when scientists think these creatures lived. Remember, not all Christians agree with those dates, but we can still learn fascinating things about God's creativity."

After controversial lessons: "What did you learn about dinosaurs today? That's interesting! You know, Christians who study dinosaurs have different ideas about when they lived. What matters most is that God made them."

This approach prevents unnecessary conflict while maintaining theological integrity. Children learn to engage with mainstream science respectfully while thinking critically from a faith perspective.

With Skeptical Relatives

Family gatherings can produce challenging conversations:

Uncle: "I suppose you're teaching little Johnny that dinosaurs lived with Noah?" Response: "We're teaching him that God created an amazing world including dinosaurs. Christians actually have different views about when they lived. We're more focused on helping him see God's creativity and love."

This deflects confrontation while maintaining witness. It shows that faith isn't threatened by questions and that Christians can think carefully about complex issues.

In Evangelistic Conversations

Sometimes dinosaur questions arise when sharing faith:

Skeptic: "I could never believe in a God who thinks the world is 6,000 years old when we know dinosaurs lived millions of years ago." Response: "I understand that concern. Did you know many Christians accept the scientific age of dinosaurs while believing God created them? The Bible's main message isn't about the age of rocks but about God's love for people like you and me. Would you be interested in hearing more about that?"

This removes unnecessary barriers to faith while redirecting to the Gospel's heart.

Building Confident Faith

Teach the Controversy

Rather than sheltering children from origins debates, prepare them to engage thoughtfully:

"Scientists disagree about how to interpret evidence about the past. Christians also disagree about how to understand Genesis. This might seem confusing, but it's actually exciting—there's still so much to discover! As you grow, you'll develop your own understanding. What matters now is knowing that God made everything and loves you deeply."

This approach builds resilient faith that can weather challenges. Children learn that questions are normal, disagreement isn't threatening, and faith seeks understanding.

Model Humble Confidence

Show that faith doesn't require having all answers:

"You know, I'm still learning about this myself. The Bible clearly teaches that God created everything, but godly people disagree about the details. I lean toward [share your view], but I have friends who love Jesus just as much and see it differently. What's wonderful is that we can keep learning together!"

This demonstrates that mature faith embraces mystery while holding firm convictions. It shows that Christians can disagree charitably while uniting around essentials.

The Ultimate Goal

Remember that dinosaur conversations are means, not ends. The goal isn't winning arguments about origins but pointing people to the Creator. Whether someone believes dinosaurs lived 6,000 or 66 million years ago matters far less than whether they know the God who made both dinosaurs and humans.

Keep this eternal perspective when answering questions. A person convinced about young-earth creationism but alienated from Christ has lost eternally. Someone who accepts evolution but finds salvation through your gentle witness has gained everything.

Let dinosaur questions become opportunities for demonstrating Christian thoughtfulness, kindness, and confidence in God's truth. Show that faith engages questions rather than avoiding them. Model how to hold convictions humbly while remaining open to learning.

Most importantly, let your life demonstrate that knowing the Creator personally transforms everything. When people see joy, peace, and purpose flowing from your relationship with Christ, questions about dinosaurs find proper perspective. The God who crafted Tyrannosaurus and Triceratops crafts new hearts in those who trust Him.

For Reflection

1. What dinosaur questions do you find most challenging to answer? Why?
2. How can you prepare children to engage with different views on origins?
3. What response strategies from this chapter seem most helpful for your context?

Practical Exercise

Practice responding to common questions with a friend or family member. Have them play the skeptic while you practice gentle, thoughtful answers. Notice where you feel defensive or uncertain. These areas need additional study or prayer. Remember, the goal isn't perfect answers but faithful witness.

Remember This

Blaise Pascal, the brilliant 17th-century mathematician and Christian philosopher, wrote: "Truth is so obscure in these times, and falsehood so established, that unless we love the truth, we cannot know it." Approaching dinosaur questions with love for both truth and people positions us to give answers that honor God and help others.

For Further Study

* Ham, Ken and Hodge, Bodie. Dinosaurs for Kids
* Poling, Judson. How to Talk to Your Kids about Evolution
* Samples, Kenneth. Without a Doubt: Answering the 20 Toughest Faith Questions
* Strobel, Lee. The Case for a Creator (Student Edition)

Chapter 14:

The Greatness of the Creator, Revealed in Dinosaurs

"For since the creation of the world God's invisible qualities—his eternal power and divine nature—have been clearly seen, being understood from what has been made, so that people are without excuse."
— Romans 1:20

Standing Before Titans

The sun was setting over the Royal Tyrrell Museum in Alberta, casting long shadows through the glass walls of the main gallery. Dr. Elizabeth Chen, a visiting theologian, stood alone before the mounted skeleton of Tyrannosaurus rex, tears streaming down her face. She had come to study how Christians engage with paleontology, but in this moment, academic analysis dissolved into pure worship.

"Lord," she whispered, "You thought of this. In Your infinite creativity, You imagined creatures so magnificent, so terrible, so utterly beyond human invention. And You spoke them into being." The dying light transformed the ancient bones into a cathedral of praise, each fossil a stone in creation's testimony to its Maker.

Elizabeth's experience echoes across centuries of believers encountering dinosaurs. From Mary Anning finding her first ichthyosaur to modern children gasping at museum displays, these ancient creatures evoke something primal—a recognition of power and artistry beyond human comprehension. In dinosaurs, we glimpse attributes of God that tamer creatures cannot fully reveal.

The God of Unimaginable Creativity

Beyond Human Categories

Human creativity operates within limits. We combine existing elements in new ways, but we cannot imagine something truly alien to our experience. Yet God created dinosaurs that shatter our categories. Consider Therizinosaurus with its three-foot claws used not for killing but for pulling down branches. Who but God would design a gentle giant with scythe-hands?

Or contemplate Quetzalcoatlus, a flying creature the size of a giraffe. Human engineers struggle to make small aircraft fly efficiently, yet God crafted a reptile that soared on wings spanning forty feet. Every calculation says it shouldn't work, but fossil trackways prove it walked on folded wings and launched into flight. God's engineering transcends our physics textbooks.

The sheer diversity astounds. Over 1,000 dinosaur species identified, with new discoveries monthly. Long-necked sauropods, armor-plated ankylosaurs, dome-headed pachycephalosaurs, sailing-backed spinosaurs—each group contains dozens of variations on themes we never would have conceived. God's creativity knows no bounds, no exhaustion, no repetition from lack of ideas.

The Artistry of Function

Dinosaurs reveal a Creator who delights in beauty within function. Parasaurolophus's elaborate crest wasn't mere decoration but a resonating chamber, allowing herds to communicate through haunting songs we'll never hear. Form and function merge in divine artistry.

Consider the tail clubs of ankylosaurs—weapons, yes, but sculpted with aesthetic grace. Or ceratopsian frills, varying from Triceratops's solid shield to Styracosaurus's crown of spikes to Kosmoceratops's elaborate curved horns. Each design solves defensive needs while displaying unique beauty. God could have made one effective frill design; instead, He crafted dozens, each a masterpiece.

Even in microscopic details, divine artistry appears. Fossilized skin impressions reveal intricate patterns—scales arranged in rosettes, tubercles positioned for flexibility, color patterns preserved in melanosomes. God attended to details no human would see for millions of years, simply because He delights in excellence.

Creative Excess

Why create Dracorex hogwartsia with its dragon-like skull? Why give Amargasaurus two parallel rows of spines along its neck? Why design Yi qi with bat-like membrane wings when feathered wings already existed? These creatures display what we might call creative excess—design beyond necessity, artistry for its own sake.

This excess reflects the God revealed in Scripture, who creates not from need but from love, not for utility but for joy. The same God who knows every sparrow designed Microraptor with iridescent feathers that shifted from black to blue in sunlight. The same God who clothes lilies in splendor gave Borealopelta countershaded camouflage patterns preserved perfectly in stone.

Dinosaurs teach us that God's creativity operates on scales we cannot fathom. He thinks in eons and epochs, crafting creatures for worlds we'll never see, delighting in forms that flourish briefly before passing away. His creative resources are infinite, His imagination boundless, His artistry inexhaustible.

The God of Sovereign Power

Power Over Time

Whether dinosaurs lived thousands or millions of years ago, their existence demonstrates God's sovereignty over time itself. He is not bound by human lifespans or attention spans. He can craft stories spanning ages, work purposes through epochs, achieve goals across geological time.

Consider the progression from early archosaurs to diverse dinosaur groups to modern birds. Whether through special creation or guided evolution, this narrative arc displays patient purposefulness. God doesn't rush. He unfolds plans through whatever timeframe serves His purposes. Dinosaurs teach us that the eternal God thinks eternally.

The fossil record, however interpreted, shows divine power over preservation itself. What are the odds that soft tissue would survive in T. rex bones? That Scipionyx would fossilize with intestines intact? That Chinese volcanic ash would preserve feathers in microscopic detail? God sovereignly ensured we would discover these testimonies to His ancient works.

Power Over Life and Death

Dinosaur extinction, whether by flood, asteroid, or gradual change, demonstrates God's absolute sovereignty over life and death. He who spoke dinosaurs into existence could speak them out again. The Creator reserves the right to end what He began, to close chapters and open new ones.

Yet this power isn't arbitrary or cruel. Each extinction cleared ecological space for new creation. Dinosaur death enabled mammalian life, eventually permitting human existence. What seemed like ending was preparation. God's sovereignty encompasses both creation and destruction in service of ultimate purposes.

The pattern continues throughout Scripture. The flood destroyed but preserved. The exile scattered but purified. The cross killed but resurrected. In dinosaur fossils, we see this deep pattern written in stone—life through death, creation through catastrophe, purpose through apparent destruction.

Power Over Process

However God created dinosaurs—instantaneously or through gradual processes—their existence demonstrates mastery over means. An omnipotent God isn't limited to one method. He can speak worlds into existence or guide billion-year developments. His power transcends our categories of natural and supernatural.

This liberates us from anxious debates about mechanisms. Whether God said "Let there be Brachiosaurus" or guided mutations and selections toward sauropod emergence, the result declares His glory. Whether radioactive decay rates remained constant or accelerated during creation week, God's sovereignty remains absolute. Dinosaurs remind us that how God works matters less than that He works.

The God of Incomprehensible Wisdom

Ecological Orchestration

Dinosaur ecosystems reveal staggering orchestration. In Morrison Formation deposits, we find evidence of complex food webs—Allosaurus hunting Stegosaurus, Ceratosaurus scavenging carcasses, Diplodocus processing tons of vegetation, Ornitholestes catching small prey. Each creature filled a precise niche in balanced ecosystems.

This balance required incomprehensible foresight. How did predator-prey ratios maintain stability? How did plant communities support massive herbivores without depleting? How did migration patterns prevent overgrazing? The wisdom required to design self-sustaining ecosystems across continents and centuries surpasses human comprehension.

Modern ecosystems pale in complexity compared to Mesozoic communities. The Dinosaur Park Formation in Alberta preserves evidence of at least 40 dinosaur species coexisting—mega-herbivores, small omnivores, various-sized carnivores, each with specific habitat preferences and feeding strategies. Orchestrating such diversity requires wisdom beyond calculation.

Information Systems

The genetic information required to build a dinosaur staggers comprehension. Consider Argentinosaurus—how do you encode instructions for growing a 100-ton animal? The bone density gradients, the cardiovascular adaptations, the metabolic requirements, the developmental timing—all specified in molecular code.

Where did this information originate? Random mutations cannot generate the coordinated instructions for building a sauropod neck that reaches 50 feet high while maintaining blood flow to the brain. The wisdom embedded in dinosaur DNA, glimpsed through protein fragments in fossils, points to a Mind transcending material processes.

Even extinct information systems display divine wisdom. Stegosaurus plates served multiple functions—thermoregulation, display, defense—requiring integrated design. The neural programming for Maiasaura parental care encoded complex behaviors. Each dinosaur embodied encyclopedias of information, testimonies to omniscient wisdom.

Purposeful Progression

The history of dinosaurs shows purposeful progression, not random wandering. Early dinosaurs were small, bipedal creatures. Through time (however long), they diversified into myriad forms filling every available niche. This isn't directionless change but orchestrated development toward goals.

The progression from dinosaurs to birds illustrates divine wisdom working through transformation. Features that seemed restrictive—bipedal stance, hollow bones, high metabolism—became foundations for flight. God's wisdom encompasses past, present, and future, designing features for purposes not yet realized.

Even extinction served wise purposes. Dinosaur domination prevented mammalian radiation. Their removal opened possibilities for creatures eventually bearing God's image. Divine wisdom orchestrates endings as carefully as beginnings, destructions as purposefully as creations.

Responding in Worship

Awe and Wonder

Dinosaurs should evoke worship. When we contemplate Quetzalcoatlus soaring over ancient seas or Argentinosaurus shaking the earth with each step, how can we not praise their Maker? These creatures display divine attributes on scales that humble human pride and inspire reverent awe.

Let museum visits become pilgrimages. Let fossil discoveries prompt doxology. Let debates about origins yield to shared wonder at the God who imagined such creatures. Whether young-earth creationist or theistic evolutionist, all believers can unite in worshipping the Creator whose power and wisdom dinosaurs declare.

Children understand this instinctively. Watch their faces encountering dinosaur skeletons—pure wonder unmarred by ideological disputes. They see what theologians sometimes miss: these bones preach sermons about a God of infinite creativity and power. May we recover such childlike worship!

Humility and Trust

Dinosaurs humble human pretensions. We who struggle to preserve species for decades encounter creatures that thrived for millions of years. We who debate their origins and extinction face mysteries beyond resolution. Dinosaurs teach intellectual humility—we know less than we think.

This humility enables trust. The God who orchestrated dinosaur history can be trusted with human history. The Creator who guided T. rex and Triceratops through their appointed times guides us through ours. His purposes may be inscrutable, but His character remains trustworthy.

When overwhelmed by life's complexities, remember dinosaurs. God managed countless species across continents and ages. Surely He can manage your circumstances. The fossil record testifies to a God engaged with creation's details while accomplishing eternal purposes.

Anticipation of New Creation

Dinosaurs point beyond themselves to new creation. Their fossils preserve beauty and power from a lost world, raising hopes for worlds to come. If God lavished such creativity on creatures destined for extinction, what wonders await in eternal creation?

Perhaps new creation will reveal what dinosaurs were always meant to become. C.S. Lewis imagined heaven containing the "real" versions of earthly things. Might glorified Tyrannosaurus transcend fearsome predator to become something unimaginably magnificent? Will Pteranodon soar in skies unlimited by atmosphere?

More certainly, the God who created dinosaurs will create wonders surpassing imagination. Eye has not seen, ear has not heard what God prepares for those who love Him. Dinosaurs merely hint at divine creative capacity. The best is yet to come.

Final Reflections

Unity in Wonder

This book has explored how Christians understand dinosaurs differently—young earth, old earth, progressive creation, theistic evolution, intelligent design. These differences matter and deserve thoughtful engagement. But they mustn't obscure shared wonder at the God dinosaurs reveal.

Imagine believers from all perspectives standing together before Argentinosaurus. The young-earth creationist marvels at God's power displayed in recent creation. The old-earth advocate sees patient purpose across deep time. The theistic evolutionist glimpses divine creativity working through natural processes. All worship the same Creator, revealed through the same bones.

Our divisions often reflect human limitations more than divine contradictions. God's truth encompasses more than any single perspective captures. Dinosaurs remind us that reality exceeds our categories, that God transcends our theological systems, that creation declares glory beyond human comprehension.

The Greater Story

Dinosaurs are not the main story but part of a greater narrative. They prepared the stage for image-bearers who could know their Maker personally. Their rise and fall foreshadowed the cosmic drama of creation, fall, redemption, and restoration centering on Christ.

In God's economy, nothing is wasted. Every Microraptor that died, every Parasaurolophus that sang, every T. rex that hunted played its role in the story culminating at Calvary. Creation groans toward redemption, and dinosaurs participated in that groaning.

Their fossils now serve new purposes—inspiring wonder, provoking questions, driving us to seek the God who crafted such marvels. In museums and textbooks, these ancient creatures continue testifying to their Maker. Their silence speaks; their absence witnesses; their bones preach.

The God Who Delights in Dinosaurs

Scripture reveals a God who delights in His works. He rejoices over creation like an artist over masterpieces. Dinosaurs show us a playful God who invented creatures beyond necessity, a generous God who created abundance destined for obscurity, a patient God who prepared Earth's story through ages.

This God invites us into His delight. He shares His creative joy with creatures capable of appreciation. Every child enchanted by dinosaurs participates in divine pleasure. Every scientist uncovering new species shares God's joy of discovery. Every believer worshipping the Creator through His works enters divine delight.

The same God who conceived Spinosaurus conceived salvation. The mind that designed Quetzalcoatlus designed redemption's plan. The power that spoke dinosaurs into existence spoke forgiveness from the cross. The wisdom orchestrating ancient ecosystems orchestrates believers' lives toward glory.

Conclusion: Living in Light of Ancient Wonders

As we close our exploration of dinosaurs and faith, what have we learned? First, that God's creation exceeds our comprehension. Dinosaurs stretch our categories and humble our certainties. They remind us we're exploring the works of an infinite Creator with finite minds.

Second, that Christians can disagree about secondary matters while uniting around primary truths. The age of Earth and mechanism of creation generate legitimate debate. But all believers affirm that God created, sustains, and will consummate His purposes for creation.

Third, that studying creation enriches faith rather than threatening it. Each dinosaur discovery reveals more of God's character. Science becomes a means of worship when pursued with reverence. Questions about origins drive us deeper into theology and doxology.

Finally, that wonder transcends argument. While this book examined various positions on dinosaurs, its goal was not winning debates but inspiring worship. The God who imagined Tyrannosaurus and Triceratops deserves our awe, trust, and praise.

Go forward with confidence. Visit museums with expectation of encountering God's glory. Study fossils as testimonies to divine creativity. Engage questions with humble confidence. Let dinosaurs do what all creation does—declare the glory of God.

The ancient dragons sleep in stone, awaiting the day when all creation will be liberated from bondage to decay. Until then, their fossils preach sermons about power, wisdom, and creativity beyond measure. May we have ears to hear what the Spirit says through these stones.

To the God who created dinosaurs, who sustains sparrows, who redeems sinners, and who will make all things new—to Him be glory forever through Christ Jesus our Lord.

Amen.

For Final Reflection

1. How has your view of God expanded through considering His creation of dinosaurs?
2. What aspects of dinosaurs most inspire worship in you?
3. How can wonder at creation lead others to seek the Creator?

A Closing Prayer

Almighty Creator, we stand in awe before the works of Your hands. In dinosaurs, we glimpse Your infinite creativity, sovereign power, and incomprehensible wisdom. Forgive us for reducing You to our small categories. Expand our vision of Your greatness.

Thank You for preserving testimonies to Your ancient works in stone. Thank You for minds capable of discovering and appreciating these wonders. Thank You most of all for revealing Yourself through Christ, the Word through whom all things were made.

Help us hold our convictions humbly, engage questions thoughtfully, and worship You wholeheartedly. May our wonder at dinosaurs lead others to wonder at their Maker. Use even these ancient bones to draw people to living faith.

Until that day when all creation is renewed and we understand fully, grant us wisdom to study, patience to learn, and hearts quick to worship. Through Jesus Christ, who holds all things together, Amen.

The Journey Continues

This book ends, but exploration continues. New fossils await discovery. Fresh insights will emerge. Debates will develop nuance. Through it all, may we remember that dinosaurs exist not as problems to solve but as wonders revealing their wonderful Creator.

Go in peace, grounded in Scripture, informed by science, and filled with wonder at the God who does all things well–from crafting ancient dragons to redeeming modern sinners. The adventure of faith and learning continues!

Appendix

These appendices provide practical resources for further exploration of the fascinating intersection between dinosaurs and Christian faith. May they serve you well in your continued journey of discovery and worship.

Major Dinosaur Groups

Overview of Classification

Dinosaurs ("terrible lizards") comprise a diverse group of reptiles that dominated Earth's terrestrial ecosystems during the Mesozoic Era. They are divided into two main orders based on hip structure:

Saurischia ("lizard-hipped") - pubis points forward

* Theropoda (bipedal carnivores and birds)
* Sauropodomorpha (long-necked herbivores)

Ornithischia ("bird-hipped") - pubis points backward

* Despite the name, birds actually descended from saurischian dinosaurs
* All ornithischians were herbivores

Major Theropod Groups

Large Carnivores

* Tyrannosaurus rex - Late Cretaceous, North America, 40 feet long
* Spinosaurus - Mid-Cretaceous, Africa, possibly 50+ feet long
* Allosaurus - Late Jurassic, North America, 28 feet long
* Carcharodontosaurus - Mid-Cretaceous, Africa, 40+ feet long

Raptors (Dromaeosaurids)

* Velociraptor - Late Cretaceous, Asia, turkey-sized
* Deinonychus - Early Cretaceous, North America, 11 feet long
* Utahraptor - Early Cretaceous, North America, 20 feet long
* Microraptor - Early Cretaceous, China, crow-sized with four wings

Early Birds and Bird-like Dinosaurs

* Archaeopteryx - Late Jurassic, Europe, first known bird
* Sinosauropteryx - Early Cretaceous, China, first feathered dinosaur discovered
* Confuciusornis - Early Cretaceous, China, early beaked bird

Major Sauropod Groups

Giant Long-Necks

* Brachiosaurus - Late Jurassic, Africa/North America, 85 feet long
* Diplodocus - Late Jurassic, North America, 90 feet long
* Apatosaurus - Late Jurassic, North America, 75 feet long

Argentinosaurus - Late Cretaceous, South America, possibly 100+ feet long

Unique Sauropods

* Amargasaurus - Early Cretaceous, South America, with double row of spines
* Nigersaurus - Mid-Cretaceous, Africa, with wide "vacuum cleaner" mouth
* Saltasaurus - Late Cretaceous, South America, armored sauropod

Major Ornithischian Groups

Ceratopsians (Horned Dinosaurs)

* Triceratops - Late Cretaceous, North America, three-horned face
* Styracosaurus - Late Cretaceous, North America, elaborate horn array
* Protoceratops - Late Cretaceous, Asia, early small ceratopsian

Hadrosaurs (Duck-billed Dinosaurs)

* Parasaurolophus - Late Cretaceous, North America, elaborate crest
* Edmontosaurus - Late Cretaceous, North America, flat-headed
* Maiasaura - Late Cretaceous, North America, "good mother lizard"

Ankylosaurs (Armored Dinosaurs)

* Ankylosaurus - Late Cretaceous, North America, tail club
* Euoplocephalus - Late Cretaceous, North America, heavily armored
* Nodosaurus - Mid-Cretaceous, North America, no tail club

Stegosaurs (Plated Dinosaurs)

* Stegosaurus - Late Jurassic, North America, double row of plates
* Kentrosaurus - Late Jurassic, Africa, plates and spikes
* Gigantspinosaurus - Late Jurassic, China, massive shoulder spikes

Pachycephalosaurs (Dome-headed Dinosaurs)

* Pachycephalosaurus - Late Cretaceous, North America, thick skull dome
* Stygimoloch - Late Cretaceous, North America, spiky head
* Dracorex - Late Cretaceous, North America, dragon-like skull

Geological Time Distribution

Triassic Period (252-201 million years ago)

* Early dinosaurs like Eoraptor, Herrerasaurus
* Small, bipedal forms
* Dinosaurs were minor components of ecosystems

Jurassic Period (201-145 million years ago)

* Rise of giant sauropods
* Diversification of theropods
* First birds (Archaeopteryx)
* Stegosaurs at their peak

Cretaceous Period (145-66 million years ago)

* Maximum dinosaur diversity
* Flowering plants appear and diversify
* Duck-billed and horned dinosaurs proliferate
* Ended with mass extinction event

Glossary of Terms

Scientific Terms

Archosaur - "Ruling reptile"; group including dinosaurs, pterosaurs, crocodiles, and birds

Biostratigraphy - Dating rocks by the fossils they contain

Clade - Group of organisms sharing a common ancestor

Coprolite - Fossilized feces providing diet information

Fossil - Preserved remains or traces of ancient organisms

Geological Column - Idealized sequence of rock layers with characteristic fossils

Gondwana - Southern supercontinent during dinosaur era

Laurasia - Northern supercontinent during dinosaur era

Mesozoic Era - "Middle life"; the Age of Dinosaurs (252-66 mya)

Morphology - Study of form and structure

Paleontology - Study of ancient life through fossils

Pangaea - Supercontinent that broke apart during dinosaur times

Phylogeny - Evolutionary relationships between organisms

Radiometric Dating - Determining age using radioactive decay

Stratigraphy - Study of rock layers

Taphonomy - Study of how organisms become fossilized

Taxonomy - Classification of organisms

Biblical/Theological Terms

Bara - Hebrew: to create (used for divine creation ex nihilo)

Behemoth - Powerful creature described in Job 40

Day-Age Theory - Interpretation of Genesis days as long periods

Ex Nihilo - "Out of nothing"; creation without pre-existing materials

Framework Hypothesis - Genesis 1 as literary framework, not chronology

Gap Theory - Temporal gap between Genesis 1:1 and 1:2

General Revelation - God's self-disclosure through nature

Hermeneutics - Principles of biblical interpretation

Imago Dei - "Image of God"; humanity's unique status

Kind (Min) - Biblical term for created categories of organisms

Leviathan - Sea creature described in Job 41 and elsewhere

Natural Theology - Learning about God through creation

Special Revelation - God's self-disclosure through Scripture

Tanninim - Hebrew: sea monsters, dragons, or large reptiles

Theodicy - Justifying God's goodness despite evil's existence

Yom - Hebrew: day (literal or figurative period)

Dating Method Terms

Carbon-14 Dating - Radiocarbon method useful to ~50,000 years

Half-life - Time for half of radioactive parent to decay

Isochron - Dating method using multiple samples

K-Ar Dating - Potassium-Argon method for volcanic rocks

K-T (K-Pg) Boundary - Rock layer marking dinosaur extinction

Radioisotope - Unstable atomic form that decays

U-Pb Dating - Uranium-Lead method for ancient rocks

Organizations and Movements

Answers in Genesis - Young-earth creationist organization

BioLogos - Organization promoting evolutionary creation

Discovery Institute - Intelligent Design think tank

Institute for Creation Research - Young-earth research organization

Reasons to Believe - Old-earth creationist organization

Resources for Further Study

Books on Different Perspectives

Young-Earth Creationist

* Morris, Henry M. The Genesis Record
* Ham, Ken. The Lie: Evolution
* Snelling, Andrew. Earth's Catastrophic Past (2 volumes)
* Wise, Kurt. Faith, Form, and Time
* Mortenson & Ury (eds). Coming to Grips with Genesis

Old-Earth Creationist

* Ross, Hugh. The Creator and the Cosmos
* Ross, Hugh. A Matter of Days
* Rana & Ross. Who Was Adam?
* Young & Stearley. The Bible, Rocks and Time
* Collins, C. John. Genesis 1-4: A Linguistic, Literary, and Theological Commentary

Intelligent Design

* Behe, Michael. Darwin's Black Box
* Meyer, Stephen. Darwin's Doubt
* Dembski, William. The Design Revolution
* Wells, Jonathan. Icons of Evolution
* Axe, Douglas. Undeniable

Theistic Evolution/Evolutionary Creation

* Collins, Francis. The Language of God
* Lamoureux, Denis. Evolutionary Creation
* Haarsma, Deborah & Loren. Origins
* McGrath, Alister. A Fine-Tuned Universe
* Venema & McKnight. Adam and the Genome

Dinosaur Science Books

General Paleontology

* Fastovsky & Weishampel. Dinosaurs: A Concise Natural History
* Brusatte, Steve. The Rise and Fall of the Dinosaurs
* Paul, Gregory S. The Princeton Field Guide to Dinosaurs
* Holtz, Thomas. Dinosaurs: The Most Complete, Up-to-Date Encyclopedia

Specific Topics

* Schweitzer, Mary. Scientific papers on soft tissue preservation
* Prum, Richard. The Evolution of Beauty
* Bakker, Robert. The Dinosaur Heresies
* Horner & Gorman. How to Build a Dinosaur

Websites and Online Resources

Young-Earth Perspective

* Creation.com (Creation Ministries International)
* AnswersInGenesis.org
* ICR.org (Institute for Creation Research)

Old-Earth Perspective

* Reasons.org (Reasons to Believe)
* OldEarthMinistries.org
* GodAndScience.org

Intelligent Design

* Discovery.org/id (Center for Science and Culture)
* UncommonDescent.com
* Darwin'sDoubt.com

Theistic Evolution

* BioLogos.org
* ASA3.org (American Scientific Affiliation)

* FaradayInstitute.org

General Paleontology

* UCMP.Berkeley.edu (UC Museum of Paleontology)
* NHM.ac.uk (Natural History Museum, London)
* AMNH.org (American Museum of Natural History)
* Smithsonian Natural History Museum

Museums and Educational Centers

Creationist Museums

* Creation Museum (Petersburg, Kentucky)
* Ark Encounter (Williamstown, Kentucky)
* Creation Evidence Museum (Glen Rose, Texas)
* Discovery Center (Branson, Missouri)

Major Natural History Museums

* American Museum of Natural History (New York)
* Field Museum (Chicago)
* Smithsonian National Museum of Natural History (Washington, D.C.)
* Royal Tyrrell Museum (Alberta, Canada)
* Natural History Museum (London)

Documentaries and Media

Various Perspectives

* "Is Genesis History?" (young-earth)
* "The Privileged Planet" (intelligent design)
* "Test of Faith" (theistic evolution)
* "Walking with Dinosaurs" (BBC, mainstream science)
* "Dinosaur 13" (story of Sue the T. rex)

Questions for Group Discussions

Chapter 1: When Christians Discovered Dragons

1. How did early Christian paleontologists like Buckland and Owen reconcile their faith with their fossil discoveries?
2. What can we learn from how Victorian Christians initially responded to dinosaur discoveries?
3. How might the history of Christian engagement with paleontology inform current debates?

Chapter 2: What We Really Know About Dinosaurs

1. Which paleontological methods do you find most convincing? Which raise questions for you?
2. How do we distinguish between observational science and historical science?
3. What role should scientific consensus play in Christian thinking about origins?

Chapter 3: Dragons in the Bible and Ancient Cultures

1. Do you think biblical passages like Job 40-41 describe dinosaurs? Why or why not?
2. How should Christians evaluate claims about human-dinosaur coexistence?
3. What's the difference between poetic imagery and historical description in Scripture?

Chapter 4: Young-Earth Creationism

1. What do you find most compelling about young-earth arguments? Most challenging?
2. How important is the age of the earth to essential Christian doctrine?
3. How can Christians who disagree about earth's age maintain unity?

Chapter 5: Old-Earth Models

1. Which old-earth model (Gap Theory, Day-Age, Progressive Creation, Framework) seems most plausible?

2. How do we handle the theological challenge of death before the Fall?
3. Can Genesis accommodate long ages without compromising biblical authority?

Chapter 6: Intelligent Design

1. Do you find the concept of irreducible complexity convincing? Why or why not?
2. Should ID be considered science or philosophy?
3. How does detecting design in nature affect your faith?

Chapter 7: Theistic Evolution

1. Can God's sovereignty be maintained if He works through apparently random processes?
2. How do theistic evolutionists handle the question of human uniqueness?
3. What are the pastoral implications of accepting or rejecting evolution?

Chapter 8: Natural Evil and Predation

1. How do you personally reconcile predatory dinosaurs with God's goodness?
2. Which theodicy regarding natural evil do you find most helpful?
3. Does animal suffering before human sin challenge your understanding of God?

Chapter 9: Time and Dating

1. What evidence regarding Earth's age do you find most compelling?
2. How should Christians handle scientific claims that seem to conflict with Scripture?
3. Is it possible to be genuinely uncertain about the age of the earth while maintaining strong faith?

Chapter 10: Extinction Events

1. How does mass extinction fit with your understanding of God's providence?
2. Do you see the dinosaurs' extinction as random, purposeful, or some-

thing else?

3. What can extinction events teach us about God's ways?

Chapter 11: Feathered Dinosaurs

1. How do transitional forms like feathered dinosaurs challenge or confirm your view of origins?
2. What does the dinosaur-to-bird transition suggest about how God creates?
3. How should classification changes (birds as dinosaurs) affect biblical interpretation?

Chapter 12: Christian Paleontologists

1. Which Christian paleontologist's approach most resonates with you? Why?
2. How can Christians in science maintain both professional excellence and faith integrity?
3. What can the broader Church learn from these scientists?

Chapter 13: Practical Responses

1. How would you answer a child's question about dinosaurs and the Bible?
2. What response strategies from this chapter seem most helpful?
3. How can we discuss origins without creating unnecessary barriers to faith?

Chapter 14: Worship and Wonder

1. How has learning about dinosaurs affected your view of God?
2. What aspects of dinosaurs most inspire worship for you?
3. How can wonder at creation lead others to the Creator?

General Discussion Questions

For Youth Groups

* If you could ask God one question about dinosaurs, what would it be?
* How do dinosaurs show God's creativity?

* What's your favorite dinosaur and what does it teach about God?

For Adult Studies

* How does your view of dinosaurs relate to your overall approach to Scripture?
* What role should scientific evidence play in biblical interpretation?
* How can churches address origins questions without dividing?

For Family Discussions

* What questions about dinosaurs do your children ask?
* How can families explore museums from a faith perspective?
* What resources have you found helpful for discussing science and faith?

It is with a sense of shared journey and profound gratitude that I welcome you to explore the life, mind, and enduring legacy of the Apostle Paul through these writings. For me, delving into Paul's world—his dramatic transformation, his complex theology, his tireless ministry, and the powerful letters he left behind—has been a deeply personal and intellectually invigorating pursuit. Paul remains a figure of immense inspiration, a "chosen vessel" whose encounter with the divine fundamentally altered not only his own identity but the course of history.

In my books, whether I am attempting a creative reconstruction like a hypothetical "final epistle" or lost letters, analyzing the psychological depths of his conversion, or seeking to illuminate the core truths within his canonical writings like Romans or the shorter epistles, my aim has consistently been twofold. First, to engage rigorously with the historical context and theological substance of Paul's message. Second, and perhaps more importantly, to bridge the gap between his ancient world and our own, making his insights accessible and demonstrating their relevance for our faith, our communities, and our personal journeys of transformation today.

I do not see these texts merely as academic subjects, but as living words carrying the potential for spiritual growth and a deeper understanding of God's grace. My hope is that as you read, you will be not only informed but also challenged, inspired, and drawn into your own meaningful encounter with the essential Paul. Thank you for joining me in this exploration. May it enrich your understanding and encourage you on your own path of faith.

Sincerely, Arthur A. Tiger

arthurtigerbooks.com

what if
Paul
wrote
today
A Final Epistle for All Believers
Based on the Original Work
If Paul Wrote a Final Epistle:
A Theological Experiment
ARTHUR A. TIGER

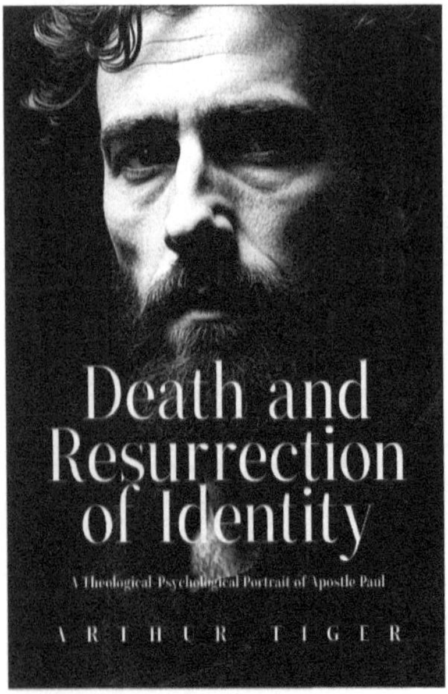

The book is a theological-psychological portrait of the Apostle Paul, exploring his radical personality transformation.

This guided prayer journal helps beginners develop a meaningful prayer life through conversation with God.

Human knowledge is limited, opening the way to faith and understanding realities beyond the senses.

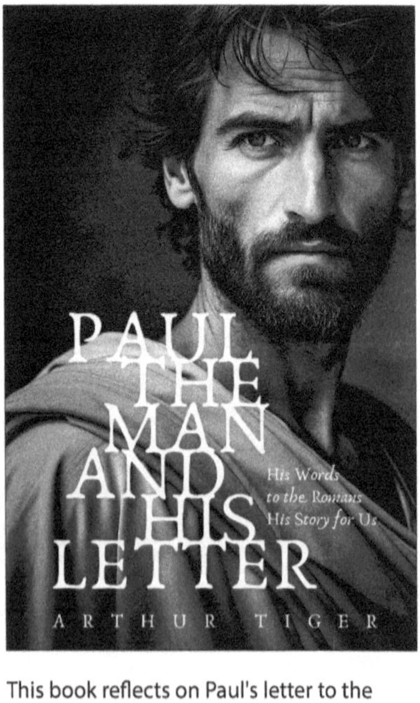

This book reflects on Paul's letter to the Romans, exploring how his words can transform lives and deepen understanding.

Theological Explorations

"The Lost Epistles of Paul" presents a bold exploration of the "unknown Paul" – reconstructing lost letters mentioned in the New Testament, uncovering theological reflections implicit in his thought, and imagining epistles the apostle might have written to various recipients.

Theological Explorations

Contemporary Christian Thought for Spiritual Growth

A theological exploration of Christian faith through historical, psychological, and biblical perspectives, designed to deepen spiritual understanding and inspire authentic Christian living in today's world.

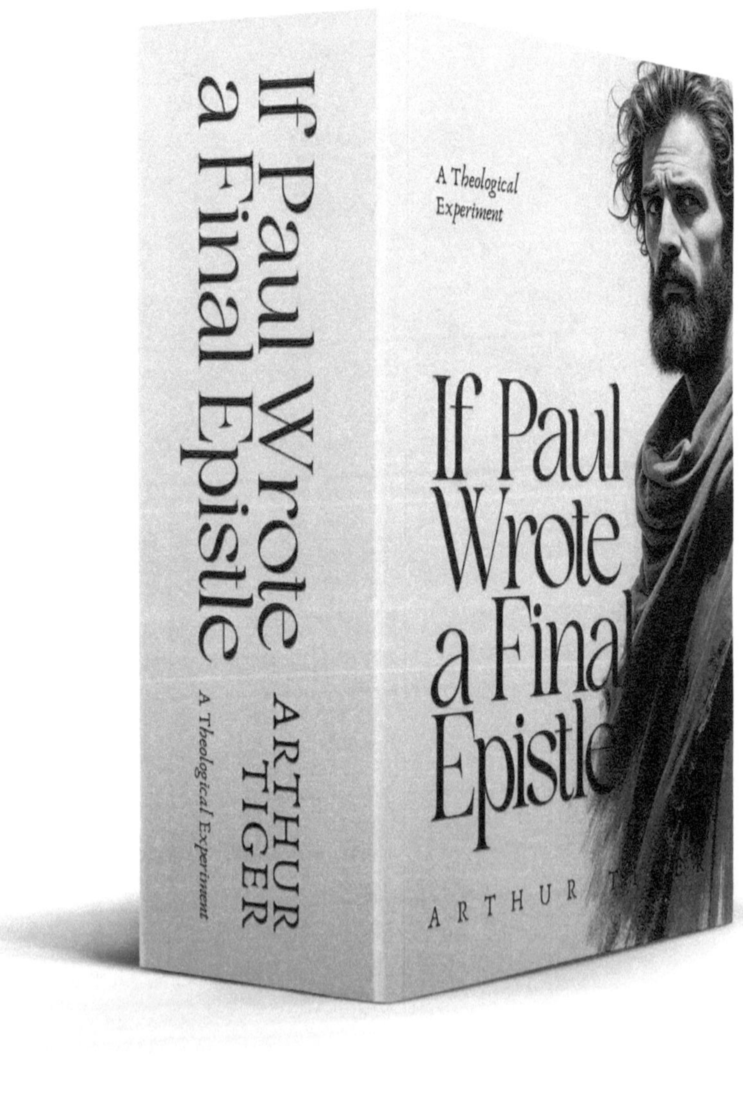

The book explores escapism and how to balance healthy rest with avoiding reality.

The book is a philosophical manifesto about life's harsh rules, shattering illusions and revealing uncomfortable truths.

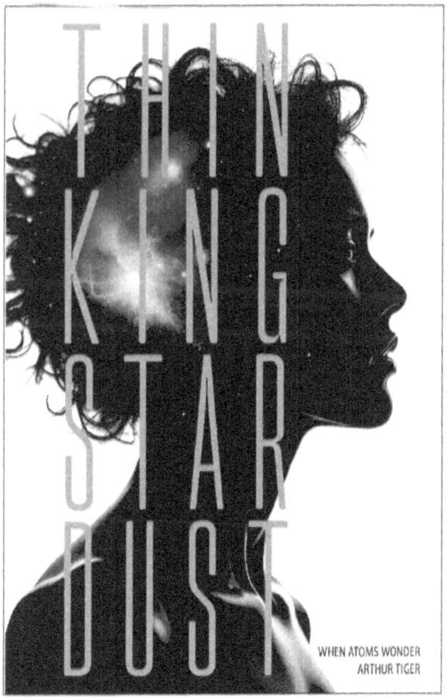

The book explores human existence, consciousness, and free will through modern physics and philosophy.

The book connects Christian faith with extraterrestrial life, asserting that science and faith are not contradictory.

Whether you're a skeptic questioning the adequacy of materialism, a disillusioned believer seeking a faith compatible with reason, or simply curious about life's deeper dimensions, this book challenges you to move beyond the artificial choice between blind faith and empty skepticism toward a more authentic engagement with reality itself.

www.ingramcontent.com/pod-product-compliance
Lightning Source LLC
Chambersburg PA
CBHW031343230926
51848CB00002B/360
9782733202920